CLEANING VALIDATION
A Practical Approach

Gil Bismuth
Shosh Neumann

CRC Press
Taylor & Francis Group
Boca Raton London New York

CRC Press is an imprint of the
Taylor & Francis Group, an **informa** business

CRC Press
Taylor & Francis Group
6000 Broken Sound Parkway NW, Suite 300
Boca Raton, FL 33487-2742

First issued in paperback 2019

© 2011 by Taylor & Francis Group, LLC
CRC Press is an imprint of Taylor & Francis Group, an Informa business

No claim to original U.S. Government works

ISBN-13: 978-1-57491-108-4 (hbk)
ISBN-13: 978-0-367-39892-7 (pbk)

A CIP record for this book is available from the British Library.

Library of Congress Cataloging-in-Publication Data available on application

**Visit the Taylor & Francis Web site at
http://www.taylorandfrancis.com**

**and the CRC Press Web site at
http://www.crcpress.com**

Contents

PART TWO
DEVELOPING A CLEANING VALIDATION PROGRAM

APPENDICES
CLEANING AND CLEANING VALIDATION PROCEDURES

Preface

Alice: "Would you tell me, please, which way I ought to go from here?"
"That depends a good deal on where you want to get to", said the cat.
"I don't much care where", said Alice.
"Then it doesn't matter which way you go", said the cat.
"So long as I get somewhere", Alice added as an explanation.
"Oh, you're sure to do that", said the cat,
"if you only walk long enough".

Lewis Carroll, *Alice in Wonderland*

Unless you have gone through the experience yourself, dealing with cleaning validation appears as an insuperable task. How can one deal with a multitude of products and a variety of equipment? What resources are needed? Where does one start? The questions seem as numerous as the products themselves, while the answers are scarce.

Although interest in cleaning validation started at the beginning of this decade, there are only regulatory recommendations and a relatively small bundle of articles providing some basic principles on selected aspects of this wide field. These are highly instructive in picturing cleaning validation as a multifaceted undertaking involving a high degree of understanding and knowledge relating to quite different disciplines, such as toxicological or clinical studies, pharmaceutics, analytical chemistry, and engineering. Yet the summation of the knowledge acquired is not to give more than a mosaic of ideas, hints, and triggers. Although very useful, it is not sufficient to build up a comprehensive cleaning validation program.

Whether or not your company has been recently inspected, or even got an observation on deficiencies of your cleaning validation program, you won't have any difficulty in convincing upper management of the need to establish a cleaning validation program in order to satisfy the regulatory authorities. The U.S. Freedom of Information Act, through the Food and Drug Administration (FDA) web site of the Internet, gives real-time access to the citations and the warning letters issued by the FDA to companies which did not adequately address this hot topic. Considering that violations cited in these dreadful letters may result in holding new applications, no company will allow itself not to enter this relatively new field of the pharmaceutical industry. Cleaning validation, however, is not only performed to satisfy authorities. The safety of patients taking your company's medicines is the prime objective, and no company would want to be involved in a litigation resulting from product contamination.

Now the burden of developing a cleaning validation program is on you, the quality assurance manager or the research and development (R&D) manager or whoever in your organization is responsible for it. You have gathered all the regulatory requirements, read all the literature relating to the subject, and participated in one or more of the many conferences and courses on cleaning validation. Your mind is burgeoning with a lot of ideas and the everlasting questions What?, When?, and How? are swirling in your mind like a predator preparing to rush its prey.

Beyond the confusion, and as for any new project, a cleaning validation program is about understanding the objective, determining the approach, setting a scientifically sound policy, organizing a program, and allocating adequate resources. The aim of this book is to provide professionals in charge of the cleaning validation program with a detailed step-by-step guide to establish a comprehensive, yet manageable cleaning validation program, and with practical examples to illustrate the implementation of the proposed program. The user-friendly database formats and practical data relating to various aspects of cleaning validation also provided in this book will make it easy for anyone to rapidly start up the program.

This book shows that what may have started with confusion ends up with a clear and manageable program which, to the authors' opinion and experience, is deemed to be acceptable by the various international regulatory authorities.

Gil Bismuth
Shosh Neumann

Part One

———

Developing a Cleaning Procedure

1

Contamination Control

INTRODUCTION

The pharmaceutical industry strives to manufacture products that fit patients' needs and expectations, while satisfying the regulatory requirements. The FDA Good Manufacturing Practice (GMP) regulations, presented in the introductory section of 21 CFR Part 210, demand that a "drug meet the requirements of the act as to safety, and has the identity and strength and meet the quality and purity characteristics that it purports or is represented to possess". The regulations go on to warn that "the failure to comply with any regulation set forth in this part [210] and Parts 211 through 226 of this chapter in the manufacture, processing, packaging, or holding of a drug shall render such drug to be adulterated under section 501(a)(2)(B) of the [Federal Food, Drug and Cosmetic] Act . . ." (FDA, April 1998). As can be seen, not only the identity, the strength, the intrinsic impurities, and pharmacological activity of a drug are considered, but also its safety with regard to harmful contamination of any kind. Obviously, the patient is not supposed to absorb any substance other than the one he has been administered.

CONTAMINATION TYPES

Drug contamination types may be classified in three categories: chemical contamination, physical contamination, and microbiological contamination.

Chemical Contamination

This type of contamination, also dubbed cross contamination, results from residuals of an active ingredient or product during the chemical synthesis or during the pharmaceutical manufacturing process, and is due to carryover such as in the processing equipment or to inadequate segregation of facilities or air-handling systems. Chemical contamination is not limited to active ingredients. Residues of cleaning agents involved in the cleaning process are equally of concern. Chemical contamination raises serious concerns relating to patient safety. The term *contamination* used throughout the book refers to chemical contamination.

Physical Contamination

Mechanical impurities and extraneous matter may be innocuous and only affect the pharmaceutical elegance of some dosage forms or may be harmful, as in the case of metallic particles in ophthalmic ointments or particulate matter and fibers in parenteral dosage forms.

Microbiological Contamination

This type of contamination is also called biocontamination or bioburden. Microorganisms are ubiquitous and special care should be taken to prevent such contamination in pharmaceutical products, in particular those containing growth supporting materials and a significant amount of water. Proliferation of microorganisms may result in the destabilization of emulsions, suspensions, and creams. The microbiological contamination of parenteral preparations presents a serious health risk, as it may cause life-threatening conditions, due to the live microorganism cells as well as pyrogens, their decomposition products. The risk of microbial growth in oral solid dosage forms is minimal due to their low water content.

CONTAMINATION CONTROL

Patient safety concerns govern the design, control, and monitoring activities with the aim of minimizing, in a consistent manner, the contamination of the facilities, systems, and equipment, through properly designed processes and procedures. Figure 1.1 presents a systematic approach to contamination control.

Facilities

Subpart C of 21 CFR 211 delineates the requirements for the design and construction features of the buildings and facilities. "Separate or defined areas [are required] for the firm's operations to prevent *contamination* and mix-ups". . . . The need for "adequate control over air pressure, microorganisms, dust, humidity and temperature. . . . Air filtration systems, including prefilters and particulate matter air filters" and the measures to be taken "to control recirculation of dust from production" and "to control contaminants" are addressed in 21 CFR 211.46. A rigorous environmental monitoring program is necessary in order to ensure that the requirements are met. Such a program shall include the following parameters: temperature, relative humidity, air flow patterns—which are controlled by pressure differentials between different processing areas, air changes, and environment microbiological monitoring.

Figure 1.1. Contamination Control

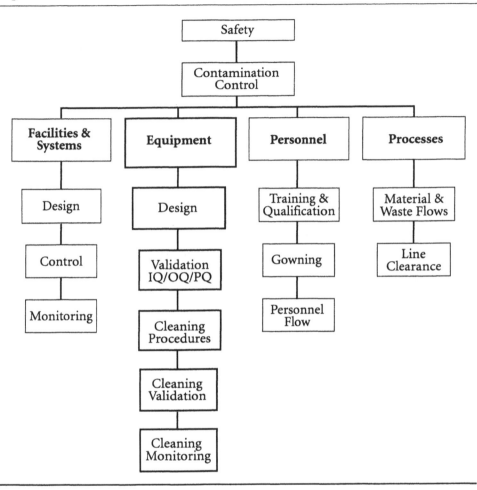

Equipment

Equipment design, construction, cleaning, and maintenance are referred to in 21 CFR 211 Subpart C. "Equipment . . . shall be of appropriate design . . . to facilitate operations . . . for its *cleaning* and maintenance". "Equipment shall be constructed so that *surfaces that contact* components, in-process materials, or drug products [the so-called product contact surfaces] shall not be *reactive, additive,* or *absorptive* so as to alter the safety, identity, strength, quality, or purity of the drug product, beyond the official or other established requirements". Also required are "written procedures . . . for *cleaning* and maintenance of equipment . . . to prevent malfunctions or *contamination*". Equipment design and construction are essential for its "cleanability" while close systems are needed for the minimization of production dust.

Personnel

The role of personnel in the minimization of contamination remains crucial, as the industry still relies a great deal on operators. Comprehensive and well-organized training sessions covering the general behavior in the plant together with specific training and qualification, will bring all the personnel to the same level of basic understanding of the effect of an individual on product quality, including the risk of contamination. Unlike other activities, which may be performed once, training to such sensitive issues and increasing personnel awareness are an ongoing process which demands creativity, changing formats and contents, and high skills.

Another aspect related to personnel is the appropriate gowning worn by operators in various locations of the plant. Company gowning, cleaned and maintained according to pharmaceutical needs, brings the personnel to the same cleanliness level and ensures that products are exposed to the desired level of cleanliness. Gowning also protects personnel from the environment, which can be harmful. Training sessions in personal hygiene and in proper gowning should, obviously, be part of the comprehensive training program. Personnel flow within the plant and its impact on contamination should be scrutinized and well designed to minimize the propagation of contamination.

Material and Waste Flows

Basic processes are designed in a pharmaceutical company to provide a behavioral pattern of the personnel to this effect. Procedures covering the handling of material and waste along the whole manufacturing process should be enforced in order to control contamination. Routine controls such as line clearance are also instituted to ensure the full separation between different products, batches, and/or campaigns.

2

Regulatory Requirements

Since 1963, the FDA has required that equipment be clean prior to use (GMP Regulations—Part 133.4). This requirement is one of the basic GMP requirements, and is mentioned in more than one section of 21 CFR 211 (FDA, April 1998). Section 211.63 relates to equipment design, size, and location and requires that "equipment used in the manufacture, processing, packing, or holding of a drug product shall be of appropriate design, adequate size, and suitably located to facilitate operations for its intended use and for its *cleaning* and maintenance". Section 211.67 further requires that "equipment and utensils shall be *cleaned*, maintained and sanitized at appropriate intervals *to prevent* malfunctions or *contamination* that would alter the safety, identity, strength, quality or purity of the drug product".

In addition to the requirements for *appropriate design* of the equipment and the need for *appropriate intervals* for cleaning, Section 211.67 introduces in detail all the parameters that should be at least included in a *written procedure* that shall be followed for cleaning. Also *equipment logs* shall be included (Section 211.182) to show the date, time, product, and lot number of each batch processed in chronological order.

The European Union (EU) and the World Health Organization (WHO) GMP guides include additional elements related to the cleaning equipment itself to be considered: "Washing and cleaning equipment should be chosen and used in order not to be a source of contamination", meaning that the cleaning agent, tools, washing solvent, and the whole procedure will ensure an effective cleaning with no residues related to the cleaning itself (EC Guide to Good Manufacturing Practice for Medicinal Products 1997; WHO Good Manufacturing Practices for Medicinal Products 1992).

Historically, FDA investigators have looked for gross contamination due to inadequate cleaning and maintenance of equipment and/or poor dust control. Also penicillin contamination in non-penicillin drug products, and potent steroids or hormones contaminating a drug product were a great concern. Indeed, a number of products have been recalled due to actual or potential penicillin cross contamination.

A few events increased awareness and pushed FDA to publish new guidelines demanding proper cleaning procedures and later, in the 90s, even cleaning validation. In 1988, Cholestyramine Resin USP was recalled due to pesticide cross contamination (FDA, July 1993; "The Gold Sheet", May 1993). The bulk pharmaceutical chemical was contaminated with low levels of agricultural pesticides while being produced. This happened while using recovered solvent drums, which were used also for recovered solvents from pesticide production process. Shipments of this contaminated bulk pharmaceutical to another facility caused pesticide contamination in the other facility.

In 1992, an import alert was instituted on a foreign bulk pharmaceutical manufacturer. The company was manufacturing potent steroid products as well as non-steroidal products in the facility using the same equipment.

In the same year, FDA officers could be quoted saying that "while the word 'validation' associated with process development has been a relatively recent buzzword, the evaluation and justification of cleaning processes have been employed in the pharmaceutical industry for many years" (FDA, May 1993). They also emphasized that, "In a July 1966 seminar on the topic of Control Procedures in Drug Production, Harley Rhodehamel discussed the development and evaluation of cleaning procedures at a large pharmaceutical manufacturing facility. The concepts and methods discussed in his paper presented in 1966 are still current today".

In the early 1990s, as previously mentioned, the FDA and other regulatory agencies started to require validation of cleaning procedures. Advances in analytical technology allowed for detection of residues at very low levels. Therefore, the requirement based on "visibly clean" for the equipment, which was the basis in the GMP, has changed to validation requirement based on quantitative determination of residues. This tendency generated some regulatory documents, including the FDA Guide to Inspection of Bulk Pharmaceutical Chemicals (FDA, September 1991). This document clearly states that "cleaning of multiple use equipment is an area where *validation* must be carried out". The issues related to cleaning validation were briefly addressed: "Cleaning process will remove residues to an *acceptable level*". A *sampling plan* was also required, while the only method mentioned was the final rinse water or solvent. The need for an *analytical method* was raised to enable the monitoring of a specific residue substance at a rational level in a practical, achievable, and verifiable manner. The nonuniformity nature of the residue was briefly discussed in relation to swab sampling.

Following the USA vs. Barr court decision reached in February 1993, more regulatory documents were published discussing cleaning validation principles and, since then, FDA investigators check more carefully for equipment cleaning procedures and cleaning validation programs (FDA CGMP Regulations, February 1993).

The court decision (USA vs. Barr) actually instituted the basic requirement that "in order for the cleaning rules to be effective, the specific methods chosen must be shown to be effective". The cleaning validation studies have to cover *all* equipment including milling machines, not just major equipment. Surprisingly, in the absence of problems, a one-run validation was suggested to be sufficient. This suggestion was dropped later on. The court decision opened the cleaning validation issue to discussions within the industry, as well as to a dialogue with FDA representatives in professional forums. This process seems to be fruitful, due to the clarifications made along the way leading to a better understanding of the considerations to be taken into account in cleaning validation studies.

Warning letters were increasingly sent to firms, including parenteral firms, repackers, over-the-counter manufacturers, and biotech firms. Parenteral firms received warning letters in 1992 for having "no documented validation to show that the cleaning procedures used for equipment have been validated" ("The Gold Sheet", May 1993).

In February 1993, warning letters were issued by FDA to three repackers. Lack of basic housekeeping (for example, cleaning procedures) was a problem observed by the agency's investigators for these operations.

Warning letters and citations (FDA 483) were issued for several biotech companies as well. FDA concerns with cleaning procedures for biological products were similarly apparent in pre-approval as well as in post-approval inspections at biotech plants ("The Gold Sheet", May 1993).

The USA vs. Barr Laboratories case has strengthened the need to enforce cleaning validation standards. The first regulatory document dealing solely with cleaning validation was FDA Cleaning Validation Mid Atlantic Region Inspection Guide published in May 1993 (FDA, May 1993). This guide actually sets the basic requirements for cleaning validation as they are known today. In July 1993, a more elaborate guide was published, Guide to Inspections of Validation of Cleaning Processes (FDA, July 1993). This guide is intended to cover equipment cleaning for chemical residues, although it is mentioned that "microbiological aspects of equipment cleaning should be considered". It is recognized that there may be more than one way to validate a cleaning process. Therefore, the guide only addresses the aspects that should be taken into consideration while validating cleaning processes. Some common validation elements for cleaning validation studies are described which are, in fact, in the same spirit as required for every validation as described in Validation Documentation Inspection Guide—FDA, 1993, and summarized as follows:

a. Written cleaning procedures should be established. Attention should be addressed to dedicate certain equipment to a specific product such as fluid bed dryer bags, and to residues originating from the cleaning detergents or solvents themselves.

b. Procedures on *how* validation will be performed should be in place.

c. *Who* is responsible for performing and approving the study.

d. Acceptance criteria should be set.

e. Procedures dealing with the subject of *when* revalidation is required, should be in place.

f. Validation protocols should be prepared before each validation study stating issues such as sampling procedure and analytical methods.

g. Study should be conducted according to the protocol.

h. Approved report should state the validity of the cleaning process.

Great emphasis is put on the elements that are to be considered while evaluating the effectiveness of the cleaning process. The first one is related to the determination of what cleaning process will be used. This decision should take into account the design of the equipment "particularly in those large systems that may employ semi-automatic or fully automatic clean-in-place (CIP) systems, since they represent significant concern".

The second element, which is described in length, is the need for a written cleaning procedure. The documented cleaning processes ought to be specific and detailed.

The third element discussed is the analytical method used to detect residuals or contaminants. The method has to be specific and sensitive enough to be able to detect the level of residues in relation to the limits which should be set for the validation. "The firm should challenge the analytical method in combination with the sampling method" which is the determination of recovery due to the sampling from the equipment surface. It is emphasized that the analytical method has to be specific and "check to see that a direct measurement of the residue or contaminant is performed".

The fourth element to be considered is the sampling technique. Two types of sampling techniques have been found acceptable. The preferred type is the direct method of sampling the surface of the equipment. The other technique is the use of rinse solutions. The advantages of the direct sampling are that it is possible to sample hard-to-clean locations in the equipment and even insoluble materials can be sampled by physical removal. In the case of rinse samples, however, insoluble contaminants or residues may be occluded in the equipment. A third method, which is less recommended, is the use of placebo product. In this case, a placebo batch is manufactured following the cleaning process and a sample of the placebo batch is tested for residues. The greatest concern using this method is that "one cannot assure that the contaminate will be uniformly distributed throughout the system".

As required for each kind of validation, the basic step toward cleaning validation is to set acceptance limits to determine whether a cleaning process is validated. Since this guide is applicable to a wide range of activities, equipment, and products, the FDA stated that it has no intention to set specifications. Therefore, it is the obligation of the company to use a rationale scientifically justifiable to set the limits. The guide brings some limits used by the industry, such as: 10 ppm, biological activity of 1/1000 of the normal therapeutic dose and organoleptic levels as no visible residue. Detergent residues should also be checked during validation. Finally, it is stated that it is usually not considered acceptable to "test until clean".

In 1995, it was reported in the literature, for example, that "The agency (FDA) wants a detailed protocol listing equipment to be sampled, how and where samples will be taken, drugs made with that equipment and (company's) schedule for sampling and study completion" (Washington Drug Letter, February 1995). In the same year the FDA issued a 483 report to another company "that cleaning procedures for dispensing, granulation and tabletting facilities were not properly validated. Validation was done after certain batches, but should have been planned also for later in production" (Washington Drug Letter, March 1995).

In 1996, it was acknowledged that FDA has increased its attention to foreign facilities. The basic requirement for adequate written cleaning procedures is still held. "Plant operators must document their cleaning methods and, sometimes even more importantly, be able to justify their approach" (Washington Drug Letter, January 1996).

In recent years, as regulatory international activities are covering wider areas, it can be seen that guidelines issued by FDA, European, and other regulatory bodies are more uniform.

In January 1996, the Pharmaceutical Inspection Convention (PIC) published a document, *Principles of Qualification and Validation in Pharmaceutical Manufacture,* which includes cleaning validation as one of the four validation areas (Document PH 1999). Basically, their recommendations are in line with the FDA requirements (FDA, July 1993). However, more clarifications and focuses are added in this document as follows:

- While cleaning validation is intended for procedures for product contact surfaces only, "consideration should be given to noncontact parts into which product may migrate".

- The term "bracketing" is mentioned here for validation. "It is considered acceptable to select a representative range of similar products and processes concerned and to justify a validation program".

- At least three consecutive applications of the cleaning procedure should be performed.

- The difference in raw materials sources and characteristics should be considered while designing cleaning procedures.

- Revalidation in cases of changes in equipment, product, or process, and also periodic revalidation should be considered.

- Reassessing manual methods more frequently than clean-in-place (CIP) systems.

- Analytical methods used for cleaning validation must be validated.

The trend in harmonizing regulations internationally can be clearly seen in two recently published draft documents for active pharmaceutical ingredients (API) GMP guides. One document was published by the PIC-PIC/S (Pharmaceutical Inspection Convention, Pharmaceutical Inspection Co-operative Scheme) (PIC-PIC/S, September 1997), and the other was published by the FDA (FDA, March 1998).

In both documents, cleaning requirements combine the aspects published in previous regulatory documents by the CFR (FDA, April 1998), EU GMP (EC Guide to

Good Manufacturing Practice for Medicinal Products 1997), and WHO GMP (WHO Good Manufacturing Practices for Pharmaceutical Products 1992). For cleaning validation, besides all previous requirements, the term "worst-case" selection of product or scenarios is pronounced clearly in these documents.

However, even in January 1998, the FDA is still citing warning letters to pharmaceutical companies: "Currently, the firm is not performing periodic testing of equipment rinse samples for product residue". Another company was cited because "the process used to clean and disinfect aseptic processing areas and equipment has not been established. In addition, the effectiveness of the agents used has not been established".

During 1997, warning letters, summarized by "The Gold Sheet" (February 1998) show that firms continue to struggle with cleaning validation. Several firms were cited for such problems, in particular non-U.S. bulk pharmaceutical manufacturers. Table 2.1 summarizes issues related to cleaning and cleaning validation as they appear in warning letters sent to various companies during 1997–1998.

"Drug GMP Report" (June 1997) reviews some GMP consultants' opinions on how to build a cleaning validation program. "GMP consultants will tell drug makers to validate cleaning procedures based on worst-case scenario. This can include picking the most sensitive drugs, and hardest to clean equipment". Some consultants have recommended conducting a risk analysis to determine which equipment is the hardest to clean.

The basic GMP regulations require companies to ensure cleanliness of equipment. This straightforward rule is established to ensure the safety, identity, strength, quality, and purity of the drug product. Over the years, a need to enforce this regulation was raised because of cases where safety of patients was a concern, and when advances in analytical technologies could allow development of cleaning validation programs. Warning letters and investigators' reports cited to companies prove that companies are still struggling to find ways to validate and establish their cleaning procedures (Washington Drug Letter, June 1998; "GMP Trends", January 1997; "GMP Trends", November 1997; "GMP Trends", December 1997; "GMP Trends", June 1998; Inspection Monitor, June 1998; Drug GMP Report, June 1998).

The following chapters of this book present a systematic and practical approach to address the regulatory concerns related to cleaning and cleaning validation.

Table 2.1. Cleaning and Cleaning Validation Warning Letters Citations, 1997–1998

1. Equipment cleaning *procedures*
2. *Adherence* to cleaning SOP
3. Cleaning *records*
4. Cleaning of equipment at appropriate *intervals*
5. Equipment *cleaning/maintenance*
6. Equipment *cleaning*
7. Cleaning of packaging *rooms and equipment*
8. Cleaning of *repackaging* equipment
9. *Facility* cleaning
10. Aseptic area/equipment *cleaning* and disinfecting
11. Cleaning *logs*
12. Contamination *prevention*
13. Controls to prevent product *mix-ups* and *cross contamination*
14. Validation of *sterilization-in-place* or *clean-in-place* processes
15. Cleaning *validation*
16. Cleaning *validation for dedicated equipment*
17. *Evaluation of adequacy* of cleaning process
18. *Effectiveness of cleaning agents*
19. *Effectiveness of cleaning process*

3

Cleaning Basic Concepts

INTRODUCTION

Cleanliness in the pharmaceutical industry is crucial to avoid safety concerns that may result from contamination. For that reason, regulatory authorities have set requirements for the basics of cleanliness in the GMP regulations. The Food Drug & Cosmetic Act even defines that: "A drug . . . shall be deemed to be adulterated . . . if it consists . . . in part of any filthy, putrid or decomposed substance" (FDA, April 1998). The cleaning operations must ensure avoidance of adulteration of products so as to minimize the risk of cross contamination between products or carryover of degradation products. Only after cleaning procedures are established and implemented can the cleaning validation phase start to prove its consistent and reliable performance.

CLEANING MECHANISMS

Cleaning can be defined as removal of residues and contaminants. These residues and contaminants can be the products themselves manufactured in the equipment (the active and nonactive materials), residues originating from the cleaning procedure (detergents), and degradation products resulting from the cleaning process itself.

Several basic mechanisms exist to remove residues from equipment, including mechanical action, dissolution, detergency, and chemical reaction (LeBlanc et al. 1993).

Mechanical action refers to physical actions, such as brushing, scrubbing, and pressurized water to remove particulates.

Dissolution involves dissolving residues with a suitable solvent. The most common and practical solvent is water because of its advantages: water is nontoxic, cheap, does not leave residues, and is environment friendly. However, in some cases it may be preferable to use a non-aqueous solvent or a combination of both aqueous and non-aqueous solvents due to the solubility characteristics of the materials. Alkaline or acidic solvents, for example, can enhance dissolution of the materials and could be advantageous.

The third mechanism, detergency, requires the use of surfactant, usually in an aqueous system. Detergents act in four different ways: wetting agents, solubilizers, emulsifiers, and dispersants. As wetting agents or surfactants they lower the water surface tension and improve penetration. As solubilizers they break up the materials using their surfactant properties. Emulsification is similar to the solubilization mechanism and refers to the creation of emulsion droplets in fats. Dispersion refers to the property to improve suspension of the materials. Usually detergents possess all these properties which broaden their action.

The fourth mechanism includes chemical reactions, such as oxidation and hydrolysis in which the residues are chemically changed. The most common example of oxidation is the use of sodium hypochlorite (chlorine bleach) as an oxidant to break down organic materials to make them more readily removable (LeBlanc et al. 1993).

Usually, these four mechanisms will play a role in combination in one cleaning procedure. During cleaning validation, the effectiveness of these mechanisms will be challenged and checked as a whole in the cleaning procedure. However, for the sake of simplicity, the worst-case product participating in the cleaning validation study, which is selected as the reference drug, will be chosen based only on the solubility of its active drug in the cleaning solvent (such as water). As will be shown in chapter 5, the solubility of the active ingredients is selected as one of the critical factors in the design of the cleaning validation, actually ignoring the role of other cleaning mechanisms. This simple attitude is very popular because it is very complex and even impossible to evaluate and quantify the role of the combination of all these mechanisms on a product containing both active and nonactive drugs.

> Cleaning mechanisms includes mechanical action, dissolution, detergency, and chemical reactions.

LEVELS OF CLEANING

Most companies have two levels of cleaning in place, depending on the level of cleanliness to be achieved. Obviously, in the context of cross contamination, cleaning between batches of different products is more stringent than cleaning between batches of the same product manufactured in a series. The first level designated as major cleaning defines a complete procedure which will clean to below a specified level of residues and contaminants left in the equipment. In this case, validation of the effectiveness of the cleaning procedure in removing residues to the required level is mandatory. The second level, designated as minor cleaning, is only performed between batches of the same product manufactured in a series, or between different strengths of the same product when the formulations are qualitatively identical. For batches of products which contain the same active material but different excipients, a minor cleaning may not be sufficient. In case of minor cleaning, the level of cleanliness depends on the product and the process, taking into consideration the element of time limitation. Usually, an abbreviated cleaning procedure (compared to major cleaning) is used which requires washing and rinsing with purified water, leaving the equipment essentially free of any visible product (Harder 1984). However, if the equipment is used for dry processing, such as milling, blending, and so on, a dry cleaning using vacuum is preferable to avoid microbial growth. For minor cleaning, cleaning validation is not required, since cross contamination is not an issue.

The API process is different from the pharmaceutical dosage form process in that the final process is largely a purification step (Lazar 1997). Therefore, different cleaning levels are considered for early and intermediate steps and for the final step. However, the same rules as previously discussed apply to API processes. Major cleaning will be performed in cases of product changeover and minor cleaning between batches in a campaign.

Generally, in dedicated equipment the issue of carryover of one product to another product does not exist since only one product is manufactured on that particular piece of equipment and the cleaning validation study includes only the detergent residues determination. However, it is important to bear in mind that, while manufacturing batches of the same product one after the other does not cause carryover of contaminants as in multipurpose equipment, degradation products between batches might change the impurity profile. This concern leads to the issue of time limitation.

Major cleaning is required between products and minor cleaning between batches of the same product. Cleaning validation is required only for major cleaning.

TIME LIMITATIONS

While the cleaning procedure details all the elements needed for a specific piece of equipment and must be validated, and while the level of cleaning determines the stringency of the cleaning procedure, time limitation is meant to manage other considerations, such as microbiological contamination or carryover of degradants which might develop with time, or simply the ease of cleaning residues of product before they dry out on the equipment.

Time limitations have to be established in relation to cleaning between different steps in the manufacturing process, taking into account the manufacturing processes and the products.

The first time limitation which has to be considered is the lapse of time between the end of manufacture and the start of cleaning. This limitation will control the ease of cleaning before the product dries out. This aspect is strongly product dependent and will prevent the degradation of materials exposed in thin layers to the environment. For a piece of equipment which is constantly used on a day-to-day basis, establishment of this time limitation may be found to be unnecessary.

The second time limitation which has to be reflected in each cleaning procedure is the lapse of time between the cleaning and drying processes. This limitation will control the microbial contamination, since working conditions are in a non-sterile environment, microorganisms will grow rapidly in humid, warm, and nutritious conditions. Therefore, it is very important to completely dry the equipment immediately after it has been cleaned.

The third time limitation is related to the frequency of cleaning between batches of the same product manufactured in a campaign in non-dedicated equipment or between batches manufactured in dedicated equipment. As mentioned, in these cases there is no need for major cleaning since no foreign product is carried over from one batch to the other. So, theoretically, it might happen that in dedicated equipment, no cleaning at all, or only minor cleaning, will be carried out. However, from the point of view of microbiological contamination and/or degradants which might develop in the next batches from residues of batches previously manufactured, the frequency of major cleaning has to be established unless there is a justification not to do so. The draft guidance for APIs relates to this subject as follows:

> Dedicated equipment should be cleaned at *appropriate intervals* to prevent the build-up of objectionable material or microbial growth. As processing approaches the purified API, it is important to ensure that incidental carryover of contaminants or degradants between batches does not adversely impact the established impurity profile. . . . However, this does not generally apply to biological APIs, where many of the processing steps are accomplished aseptically, and where it is often necessary to clean and sterilize equipment between batches (FDA, March 1998).

The fourth time limitation is related to the need for re-cleaning cleaned equipment waiting for the next use. This cautious act usually holds for equipment which is not used very frequently. A rinse before use of the equipment will get rid of dust which might have been inadvertently introduced into the equipment.

All these time limits are seldom based on a full cleaning validation study, but rather empirically and arbitrarily fixed to be as short as feasible in a production environment. A standard operating procedure and a means for the verification of compliance to these time limitations should be in place.

For products prone to microbial proliferation or known to degrade rapidly in the wet stage or firmly adhering to the product contact surfaces after drying, cleaning must be performed immediately after processing and this should be clearly mentioned in the manufacturing procedure. When none of these considerations prevail for consecutive batches of the same product manufactured in a series or for a product manufactured in dedicated equipment, a larger and arbitrary time limit may be set and cleaning is performed at the end of the series. Information for establishing these time limitations, even though set arbitrarily, is supported by the knowledge of the processes and the products. In addition, more general information is gathered from routine testing which can further give confidence to these limits. This includes microbial count testing of purified water, used as the final rinse. Routine microbial environment monitoring should estimate the microbial contamination of the working environment. Finally, microbial limit counts of the product may give an indication on the overall handling in all different manufacturing steps. Random sampling and testing is performed on products for which microbial testing is not mandatory.

The cleaning procedure should define time limitations in relation to

a. time between end of manufacturing and start of cleaning.

b. time between final rinse and drying.

c. frequency of major cleaning for manufacturing batches of the same product in campaign.

d. time until additional cleaning is performed for unused clean equipment.

4

Cleaning Procedure

INTRODUCTION

A vital element of cleaning is the cleaning procedure which combines all the knowledge and considerations relating to the equipment design, the manufacturing process, and materials and tools used in the cleaning procedure. All this information is given to the operator who has been trained to perform the cleaning operations effectively and safely. Although the cleaning procedures have to be specific to the equipment and product, it is more practical to use a limited number of cleaning agents, the same cleaning process, and the same methodologies. This approach makes these routine procedures easy to handle and validate.

EQUIPMENT DESIGN

Points to consider while developing a cleaning procedure, and that should be considered while acquiring a new piece of equipment, relate to cleanability and include: the material of construction, design, the degree of dismantling, rinsability, and product carryover potential, originating mainly from hard-to-clean locations.

In most cases, equipment for pharmaceutical use is constructed of stainless steel. However, glass, silicone, PVC, and other materials are also used. The equipment has to be suited to the cleaning mode so the surfaces are not affected by the cleaning operation, and residues are neither generated nor transferred. In the case of a clean-in-place (CIP) system, for example, attention should be paid to several elements: the quality of all building materials, integrity and sealing of welds, isometry and draining, design of sampling points, and installation of sensors or measurement devices (Laban et al. 1997).

Equipment design should be evaluated in relation to general GMP concerns and cleaning issues as described in Table 4.1.

The most important issue related to cleanability of equipment is related to hard-to-clean locations. Usually these locations are poorly accessible sites of the machine, in which contaminants and residues can be trapped. Therefore, as part of cleaning validation study, the hard-to-clean locations must be identified and sampled, and the samples tested for levels of contaminant and residue leftovers after the cleaning process has been conducted.

Table 4.1. General GMP Concerns in the Design of Processing Equipment

---

1. Product contact surfaces shall not be reactive, additive, or absorptive. Surfaces shall be hard and smooth, and shall not shed particles. Surfaces shall withstand the repeated use of cleaning and sanitizing agents.

2. The equipment shall be designed so as to prevent dead ends or spots, in which products could accumulate and be hidden from the operator's view.

3. All parts of the equipment shall be easily dismantled to permit visual inspection.

4. Valves shall be of a sanitary type.

5. All connections to utilities shall be fitted with a back flow prevention valve.

6. Vessels and containers shall be drainable.

7. Valves and instrumentation shall be installed flush with product contact surfaces.

8. Small parts, screws, and bearings, if unavoidable, shall be fixed in such a way as to prevent their accidental fall into the product.

9. Any substance required for operation, such as lubricants and coolants, shall not come into contact with the product. The engine shall be located or designed so as to prevent accidental dripping of these substances into the product.

10. The engine, the electric and electronic parts, should be hermetically encased to allow for easy cleaning and to prevent safety concerns during cleaning.

The cleaning procedure development should consider the design of the equipment and the hard-to-clean locations that will be challenged in the cleaning validation study.

MANUFACTURING PROCESS

The cleaning procedure should be developed considering the type of process executed by the equipment. Different types of processes will leave different types of residues to be removed. Topical preparations, such as creams and ointments, can be difficult to clean due to the fact that they can contain many ingredients with different characteristics, part of them thick, oily, and fatty constituents. These materials are present in high concentration in the formulation. On the contrary, injectable aqueous solutions can be easy to clean due to a limited number of ingredients which are usually water soluble and present in relatively low concentrations. The cleaning procedure can be different mainly in the prewash step where the gross residue is removed. For creams, the use of high temperature can break the consistency of the cream and make it easier to remove. In the case of solutions, a superficial rinse can be sufficient as the prewash step.

OPERATOR

"Unless a process is fully automated, people are an important variable in a process and this variable must be taken into account if we wish to have processes that consistently and with a high degree of assurance produce products with the requisite quality characteristics expected by our customers" (Kieffer 1998). This statement is valid for cleaning processes as well as for any other manufacturing operation. The operator can be the weak link, especially in manual methods and in automated methods if human interventions of any kind are possible. The key to overcome this problem is to make sure that the operator is very well trained to perform all the cleaning procedures in a reliable way. Above all, the procedure should be developed so that the operator's health will not be in danger at work.

Training of the operator is of great significance in the cleaning process. Since many variables are inherent in the manual method originating from this type of activity and are difficult to quantify, it is impossible to instruct the operator how much force to put in the scrubbing step as much as it is difficult to make sure that all the hard-to-clean locations are reached. This is true for the same operator in different days and moods, and more significantly for different operators with different skills, attitudes, and ability. For some technologies (mixers, blenders, containers and even tablet presses), the operator's parameter can be minimized by using CIP systems, but for some manufacturing steps, it seems almost impossible at present until different concepts are developed. A multiproduct packaging line for solid oral dosage forms, for example, is generally regarded as one of the most sensitive parts in a pharmaceutical facility in the sense of cross contamination originating from products (and mix-ups from packaging materials). Clear and friendly procedures as well as good training can help minimize the above-mentioned deviations. Procedures must detail the quantitative parameters as much as possible and still be kept simple. As Kieffer summarizes: "One needs first of all well-designed, fail-safe, simple optimized processes; secondly, people qualified to do their tasks in these processes, and thirdly, an empowering, motivating environment" (Kieffer 1998). The third condition relates to the overall atmosphere and culture built into the manufacturing plant, which cannot be documented, but can serve as a driving force to perform activities with understanding and to take responsibility for them.

In CIP methods, extreme conditions such as high temperature, high pressure, and/or high or low acidity or alkalinity can be used, this cannot be done in manual methods due to operator's safety concerns. Only moderate conditions are used. Therefore, safety warnings should be included in the cleaning procedure to avoid accidents. Only when these conditions are fulfilled and reproducibility in the execution of the cleaning procedure is attained can the cleaning validation studies be initiated.

> Training of the operator and his safety should be of great concern, especially in manual cleaning.

MATERIALS AND TOOLS

Each cleaning procedure has to accurately describe the materials and tools that should be used. It is obvious that the cleaning performance is strongly dependent on these variables, therefore they should be controlled. The cleaning materials and tools will be chosen taking into consideration several issues: the type of equipment, the type of manufacturing process (cream, solution, tablet . . .), and the type of cleaning procedure (manual, automated). In a multiproduct manufacturing plant, it would be extremely wise to have a very limited list of different materials and tools. First, this facilitates the logistic management of the cleaning processes; second, it gives the operator a simple common routine to perform, which can help to attain a better reproducibility; third, from the validation standpoint, the idea of executing the validation study on one product per equipment type is applicable only when the same cleaning procedure (with the same materials and tools) is used for all products manufactured on that equipment and, economically speaking, this results in a completely different scale of investment. For certain products, if the routine procedure is not sufficient due to special residues resulting from the cleaning process itself or if colored residues can only be eliminated with special care, it is advisable to apply a pretreatment which will eliminate the specific residue and then perform the routine procedure. This kind of pretreatment or prewash (which may include the use of acid, alkaline, or oxidising agents) is usually of a chemical nature resulting in a change of the residue to another entity which can be easily cleaned.

The cleaning procedure therefore should accurately define the solvents, cleaning agents, and cleaning tools to be used.

Solvents

Water is the most common solvent used because it is inert, nontoxic, cheap, environment friendly and freely available. For the prewash step, tap water will be sufficient, while for the final rinse purified water or (for sterile preparations) water for injection is used. In special cases, organic solvents (such as alcohol) may be used, due to their solubilization properties.

Cleaning Agents

Cleaning agents can be organic solvents, acidic/alkaline materials, or detergents which can act in different mechanisms. In some cases, single component cleaning agents are used, while in others, formulated cleaners are preferred. In any case, it is important to consider the following aspects listed in Table 4.2 while choosing a cleaning agent.

Organic solvents are used based on their properties to dissolve certain products, which would be difficult to clean effectively with water. This is particularly true in API manufacturing. In some cases, they will be used because of intolerance of some machine parts to water exposure. For example, alcohol must be used to clean a tablet press, particularly the die table and other parts, to avoid corrosion. A combination of organic solvent with water may give better results from cleaning and safety perspectives.

Table 4.2. Cleaning Agent Properties

---

1. It should not be corrosive to the equipment surface.

2. It should be nontoxic.

3. In case of manual procedures, it should not be harmful to the operator.

4. It should be as friendly as possible to the environment.

5. It should be very easily dissolved in water, so it can be easily rinsed.

6. It should be able to solubilize or suspend the product so that the latter can be removed without leaving any residue.

7. In case of CIP, because of the high pressure used in the process, it should not produce foam.

Alkaline agents (sodium or potassium hydroxides) can eliminate organic stains from various sources while acidic agents (nitric, phosphoric, or citric acids) apply to mineral stains (Laban et al. 1997). These cleaners can be used in a simple formulation or a complex one containing other functional additives, such as dispersants or chelating agents, for a stronger and wider effect.

The use of a formulated detergent has advantages over the use of a single component detergent because a formulated detergent has several components, which can together act in several ways such as wetting, solubilizing, emulsifying, and dispersing. As a result, it can act on formulations of multiple components with different characteristics. "Some facilities like to have one universal cleaning product and process for all their systems. A formulated cleaner can provide the flexibility for adequate cleaning for many—although certainly not all—types of residues" (LeBlanc 1993). Due to their multiple actions, the formulated detergents are more effective in their use, so that solvents, which are discouraged for use from an environmental perspective, can be avoided.

Cleaning Tools

The WHO GMP requires that *"washing and cleaning equipment* should be chosen and used as not to be a source of contamination" (WHO Good Manufacturing Practices for Pharmaceutical Products 1992). This means that if brushes, rags, or fabrics are used they should not release hair or particles. Cross contamination should be avoided while using the same multipurpose tools. Vacuum cleaners might pose this risk if they are not properly cleaned between different products.

Control of Materials and Tools

The key issue for having reproducible cleaning procedures, in addition to the thorough training of the operators, is to have complete control over all materials and tools. This will prevent contamination of equipment, facilities, and products. Control should be established as follows:

- Water quality:
  Tap water, which is used for initial rinse, should be tested periodically for microbiological quality. Purified water, which is used for final rinse, should be of the same quality as for the corresponding manufacturing activities. Water has to be routinely chemically and microbiologically monitored.

- Solvents:
  Solvents used for rinsing should be of a defined grade and from a manufacturer which is approved by the company as a supplier of this material for this purpose. A certificate of analysis should be received from the manufacturer stating conformance with specifications. A procedure for receipt (as for raw materials), and a testing plan should be in place.

- Cleaning agents:
  It is recommended to obtain from the manufacturer the safety data sheet and, under a secrecy agreement, the composition of the cleaning agents. A procedure for receipt and testing should be in place and instructions for use should be defined in the cleaning procedure. To ensure a consistent quality, a declaration should be signed by the manufacturer not to change the composition without prior notice.

- Tools:
  Only brushes and fabrics that do not shed hair and are lint-free should be used and periodically checked.

The change control procedure should evaluate the impact of a change in solvents, cleaning materials, and tools on the cleaning procedure with respect to cleaning validation.

> Solvents, cleaning materials, and tools should be described in the cleaning procedure. Any change should be evaluated in relation to cleaning validation through the change control procedure.

CLEANING PROCEDURE

Types

Methods used for cleaning equipment can be divided into manual procedures and automated procedures, such as CIP. In all cases, cleaning procedures must prove to be effective, consistent, and reproducible.

FDA's guideline for API manufacturing recommends that: "where feasible, clean in place (CIP) should be used to clean process equipment and storage vessels", in order to reproduce exactly the same procedure each time (FDA, March 1998). With a manual procedure, one must rely on the operator's skills and a thorough training of the operator is necessary to avoid variability in performance. However, in some instances, it may be more practical to use only manual procedures.

Manual Cleaning

Most cleaning procedures follow the same pattern of cleaning sequences and the thoroughness of each step in a sequence depends on the equipment and the product (see Table 4.3).

- Disassembly:
  The degree of disassembly should be detailed in the cleaning procedure. While for certain types of equipment, such as a high-shear mixer, it will mainly include disassembly of the impeller, other types of equipment, such as a tablet press or a packaging line, will require extensive disassembly.

- Prewash:
  This step includes dedusting and removal of material and waste by several techniques. Vacuum cleaning is used when accessibility to surfaces is difficult, for example, in the case of the tablet press and the packaging line. It is important to clean the vacuum cleaners themselves to avoid cross contamination. Although not recommended, blowing with pressurized air can only be used, in conjunction with an efficient suction device, if the equipment is located in a segregated place and there is no danger of cross contamination and no risk to the operator. Prewash can also be conducted by partially filling the equipment with tap water, operating the equipment, and then draining the equipment.

Table 4.3. Manual Cleaning Sequence

1. Disassembly	—
2. Prewash	Tap Water
3. Wash	Cleaning Solution
4. Rinse	Tap Water
5. Final Rinse	Purified Water
6. Drying	Hot Air
7. Visual Inspection	—
8. Reassembly	—

- Wash:
This step is the most important step in removing all the residues from the equipment surfaces. For equipment used in a wet environment, such as containers, blenders, mixers, and fluid bed dryers, washing is executed with brushes and sponges. For equipment used in a dry environment, such as a packaging machine, a wiping technique is used. The cleaning will be performed with absorbent fabrics soaked with a detergent solution or alcohol solution as in the case of the tablet press.

- Rinse:
The initial rinse removes all contaminant residues of the product and detergent. Tap water is usually sufficient for this purpose.

- Final rinse:
The final rinse is usually done with purified water to remove cleaning agent residues left after the initial rinse. Sometimes hot water is used to facilitate water evaporation from surfaces and to allow easier drying.

- Drying:
This step is crucial to avoid microorganism growth, especially in equipment used for the preparation of liquids and semisolids. Drying can be done simply by using a dry absorbent fabric in case of equipment such as a packaging machine. Sometimes, when alcohol solution is used to avoid surface corrosion as in the case of the tablet press, compressed air can be blown (only if the machine is located in a segregated area) followed by lint free fabric or paper wipes. Heating can be used in the case of containers and double-jacketed tanks to dry equipment.

- Visual inspection:
21 CFR 211.67 requires that "inspection of equipment for cleanliness immediately before use" should be conducted. Therefore, visual inspection has to be included as part of the cleaning procedure. Even though it is a qualitative evaluation, experiments show about four micrograms of material per square centimeter on a stainless steel surface can be seen (Fourman and Mullen, 1993). This examination should be conducted before the assembly is executed to allow inspection of all hard-to-clean locations in the equipment and all the disassembled parts.

- Reassembly:
Reassembly is carried out only after conducting visual inspection following the order defined in the cleaning procedure.

Automated Cleaning

Nowadays, most pharmaceutical equipment manufacturers can provide the equipment fitted with a cleaning-in-place or a washing-in-place device. In common language, the two verbs are almost indifferently used. "To wash", according to Webster's *New World Dictionary*, is "to clean with water or other liquid", and "to clean" is "to make clean (free of dirt and impurities)". After washing or cleaning, equipment is "washed" or "cleaned", but it can be determined as "clean" only after its cleanliness has been tested. The two actions do not achieve the same degree of cleanliness.

Manufacturers of washing-in-place systems do not make any claim regarding the level of cleanliness, and usually some degree of disassembly and hand scrubbing is necessary in order to complete the cleaning. While it is advisable for consistency of the cleaning operation and for cost considerations to use automated cleaning systems, it must be kept in mind that not all automated systems can provide at any given site of the processing equipment the degree of cleanliness required to minimize cross contamination. The following example serves as an illustration of the above-mentioned fact. The large cylindric smooth surfaces of a simple piece of equipment like a twin shell blender may be effectively cleaned with hot water and detergent delivered by a high pressure spray ball, but the discharging valve of the same blender with a much smaller surface area than the cylinders has to be disassembled and hand scrubbed.

Cleaning-in-place (CIP), on the other hand, means cleaning the equipment to an acceptable, quantifiable, and validatable level of cleanliness, so that another product may be manufactured in the equipment immediately after cleaning-in-place without any human intervention. The level of cleanliness to be achieved should be agreed between the vendor and the buyer and must be guaranteed by the vendor, based on documented evidence.

For a CIP system to be effective, the most critical factors are the cleaning procedure, determined by the product and process characteristics, and the processing equipment design.

The development of a CIP procedure is similar to that of a manual cleaning procedure and depends on such factors as the solubility of active and inactive ingredients and the presence and amount of lubricating, dispersing, and wetting agents. A CIP system takes advantage of the use of cleaning conditions that could be harmful for operators: high temperature, high pressure, high detergent concentration, or high or low pH of the cleaning agents. The basic CIP variables are: time, pressure, volume, temperature, cleaning agent concentration, recycling, and the sequence of cleaning, rinsing and drying. A typical CIP cycle runs as shown in Table 4.4.

Table 4.4. CIP Sequence

1. Prewash	Tap water
2. Wash	Cleaning solution
3. Blow out	Compressed air
4. Rinse	Tap water
5. Final rinse	Purified water
6. Blow out	Compressed air
7. Drying	Hot, compressed air

A simplified scheme of a CIP system was shown in Figure 4.1.

To be able to rely on the automated cycles for cleaning purposes, it is necessary to qualify and validate the CIP. *In addition, it should be demonstrated that parameters such as pressure, temperature, and volume can be controlled and monitored, and that in case of failure, an alarm system should be in place.* As a routine practice: "once CIP systems are validated, appropriate documentation should be maintained to show that critical parameters (e.g., temperature, turbulence, cleaning agent concentration, rinse cycles) are achieved with each cleaning cycle" (FDA, March 1998).

The design criteria which should be followed when installing a CIP system are as follows:

- The equipment must be cleanable. For example, the vessels should be made of stainless steel (with no dead zones), and fully drainable. The piping also should be made of stainless steel, use sanitary valves with no sudden change of pipe diameter, and shouldn't have dead legs.

- The CIP system should be installed in a technical area, not in the production area.

- The CIP sequence should be reproducible by controlling the process parameters, such as: *temperature* of rinsing water and cleaning solutions, *time* of each step, *concentration* of the cleaning agent, *volume* of rinsing water and cleaning solutions, *flow rate*, and *mixing speed*.

- Avoid microbiological contamination by use of purified water at final rinse and heat drying.

- Avoid contamination between the CIP system and the equipment by using automated back flow prevention valves.

Documentation

The documentation of the cleaning procedure is essential. The cleaning philosophy of the company should be reflected through this document and expressed in a simple yet detailed way to the operator.

The cleaning procedure must be simple, clear, precise, and documented in a way that helps the operator to execute the cleaning in a reproducible and safe manner. The basis for relying on manual cleaning performed by an operator and sometimes by different operators, knowing that this might raise concerns related to wide variations between operators, is that the procedure is written in a way that eliminates ambiguities as much as possible. This is also the basis on which these procedures can be validated and only after validation visual inspection will be sufficient to evaluate the cleanliness of the equipment.

Figure 4.1. 1. CIP system installed with tanks, heat exchangers, detergent dosing components, and others. 2. CIP-pump to reach pressure and flow as required. 3. The equipment to be cleaned. 4. CIP-return pump—to circulate and drain the cleaning solution and water.

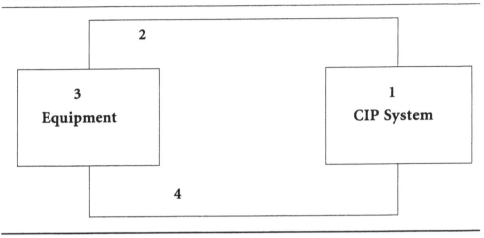

The cleaning procedure must present detailed step-by-step directions to be followed, preferably in a checklist format. The cleaning procedure shall contain all the relevant parameters and information such as volumes of detergent, solvent, temperature, time, and pressure. This information should be given in detail to enable the operator to always conduct the cleaning in a reproducible manner.

Efforts should be made to standardize all or most of the cleaning procedures. For example, when possible, all the manual procedures should be written in the same format, give the order of cleaning steps, use the same detergents, and use the same types of water or solvents in the same steps with same cleaning materials and tools. This will provide the operator a routine and easy-to-manage procedure. Sometimes, the most common procedure is not sufficient for the cleaning of a certain product. In this case, additional steps can be added to the common procedure using chemically reactive cleaning materials such as hypochlorite (chlorine bleach), when oxidative agents are needed, or alkaline or acidic solvents such as sodium or potassium hydroxide or citric acid. This additional step will be followed then by the common procedure. Only when the common procedure cannot be utilized, a different suitable procedure shall be used. The same holds for CIP procedures, the same detergents and solvents should be used for all CIP systems existing in the plant, if possible.

The documentation of the cleaning performance provides evidence of the control on this procedure. The cleaning procedures must be written and the operator has to certify, by his or her signature, the performance of the cleaning sequences to enable tracking and investigating an event related to cleanliness, and to make appropriate corrective action.

Each cleaning procedure has to include the steps outlined in the following box.

1. Objective: Cleaning of equipment and the edition number of the procedure in order to control changes, and also to be able to track them.

2. Equipment name and identification number: For similar or identical pieces of equipment, the same procedure can be used bearing all the identification numbers and the names of the equipment.

3. Cleanliness level: The same procedure can be used for minor and major cleaning. It should be indicated which steps are relevant for each level of cleaning.

4. Operator's name and his signature.

5. Precautions and safety warning should be specified.

6. Degree of disassembly has to be directed so that accessibility to hard-to-clean locations is enhanced.

7. Cleaning tools and materials should be specified. If brushes or sponges are used, it should be written in the relevant step. For cleaning agents, their name, concentration and their dilution instructions have to be part of the procedure. Also, for the water or the solvents which are used, their quantity or volume, temperature, and quality will be written.

8. Cleaning instructions should contain a step-by-step description. For CIP procedures, the sequence should set parameters for each step. Parameters such as time, temperature, pressure, volume, and flow rate of the solvent will be programmed for each step: prewash, rinse and final rinse. The operator should be made aware of how to treat hard-to-clean locations, such as: how to clean behind baffles, inside nozzles, and under the agitator. If an agitator is operated while cleaning, the agitator speed and time of operation should be mentioned.

9. Drying: This step is very important to prevent microbiological proliferation. Sometimes the final rinse is conducted with hot purified water to facilitate evaporation of the water.

10. Visual inspection: No traces or particles visible to the naked eye should be observed after the cleaning.

11. Cleaned status has to be indicated either by putting a label or a card on the clean equipment, or moving it to a clean area, or hanging a clean card on the room.

An example of a standard operating procedure for cleaning is presented in Part Three of this book.

Part Two

———

Developing a Cleaning Validation Program

5

Cleaning Validation Policy

INTRODUCTION

Now that the optimal cleaning procedure has been developed—whether manual or automated—the question remains, What do we have to demonstrate in cleaning validation? Ideally, a well-designed cleaning procedure should leave the manufacturing equipment free from residues of the previous product and a visual inspection would suffice to verify the equipment cleanliness. It has been shown, however, that visual detection has limitations and that sufficiently sensitive analytical methods may detect residues beyond the human eye capability. For dedicated equipment where cross contamination is not an issue, a visual examination after cleaning may indeed be used as the sole cleaning validation criterion. But, for multipurpose equipment, we will want to show that not more than an acceptable residual amount of any product may be found in a defined quantity of any other product, without adversely affecting patient safety. Consequently, we should define what is the acceptable residual amount of the contaminant product and how this should be determined, and what is a defined quantity of the contaminated product. Cleaning agent residues should be determined for dedicated, as well as for multipurpose equipment.

What contaminants are we looking for in cleaning validation? In most cases, a product contains a few inactive ingredients in addition to one or more active substances. The cleaning procedure is able to introduce residues related to the cleaning procedure itself, such as detergents and solvents. Degradation products and impurities can also develop in the cleaning process. To make the validation effort manageable and practical, however, it is deemed logical and acceptable to monitor residues of the active substances and of the cleaning agent in order to demonstrate the effectiveness of the cleaning procedure. The FDA's Guidance for Industry, Manufacturing, Processing or Holding of Active Pharmaceutical Ingredients, Draft for Discussion (March 1998), states that the "residue limits [for APIs] . . . should encompass raw materials, intermediates, precursors, *degradation products*, isomers and cleaning agents". However, from the reading of the FDA Guide to Inspections of Validation of Cleaning Processes (July 1993), it would appear that the requirements for degradation products do not apply to pharmaceutical dosage forms: "Unlike finished pharmaceuticals where the chemical identity of residuals are known (i.e., from actives, inactives, detergents), bulk processes may have partial reactants and unwanted by-products which may never have been chemically identified".

For an operation involving a small number of products with a limited number of pieces of equipment, cleaning validation may be addressed by validating the cleaning procedure of each piece of equipment for each product manufactured in the equipment train. For equipment used for a wide variety of products, this approach is not feasible due to the enormous analytical load involved in the development and the validation of sufficiently sensitive analytical methods. Consequently, a practical approach is to be adopted to develop a manageable program based on adequate assumptions, taking into consideration worst cases and using additional safety factors to ensure, with a high degree of confidence, that both patient safety and regulatory requirements are satisfied.

In the absence of clearly defined requirements and of authoritative publications, the industry is struggling in order to define a practical approach to make the cleaning validation effort manageable. An extensive literature search shows that different approaches have been adopted by different companies, depending on the complexity of their operations—namely the number of products manufactured and the number of pieces of equipment involved in the manufacturing processes (Laban et al. 1997; Mendenhall 1989; McCormick and Cullen 1993; McArthur and Vasilevsky 1995; Jenkins and Vanderwielen 1994; Hwang et al. 1997; Nilsen 1998; Forsyth and Haynes 1999; PDA Technical Report No. 29 1998; LeBlanc 1998; Shea et al. 1996).

By definition:

> The objective of cleaning validation is to attain documented evidence which provides a high degree of assurance that the cleaning procedure can effectively remove residues of a product and of the cleaning agent from the manufacturing equipment, to a level that does not raise patient safety concerns.

The cleaning validation program, once established, shall cover all the manufacturing operations for all products and include a tracking system to allow for the timely performance of the cleaning validation of newly developed products. Thereafter, a change control system should be in place to evaluate the impact of changes in product formulation, equipment, and/or cleaning procedure, in order to maintain the equipment cleaning validation status.

Part Two of this book focuses on the cleaning validation of equipment in complex, multiproduct pharmaceutical operations. The policy and its underlying basic principles may be applied also to active pharmaceutical ingredients (API) (Romanach et al. 1999), biological and biotechnological products (McArthur and Vasilevsky 1995; Lombardo et al. 1995; Sofer 1995; Baffi et al. 1991), with appropriate adaptations and in conformity with the applicable regulations. A cleaning validation program is presented step by step as follows:

1. Policy

2. Contamination Acceptance Limits

3. Planning and Execution

4. Monitoring and Maintenance

Entering into a cleaning validation program is a major commitment and a time-consuming and costly activity to which significant resources have to be allocated. The objective is a cleaning validation program, which covers all the manufacturing equipment and is manageable and cost effective without compromising the quality of the products. Therefore before embarking in the program, a policy should be defined in writing, agreed upon, and approved by upper management. The policy must include the basic principles covering all the aspects related to cleaning validation which, when brought together, create the manageable cleaning validation program.

POLICY

The cleaning validation policy consists of the following steps:

- Selecting the **worst case related to the product**
- Selecting the **worst case related to the equipment**
- Selecting the scientific basis for the **contamination limit**
- Selecting the **sampling method**
- Selecting the **analytical method**

SELECTING THE WORST CASE RELATED TO THE PRODUCT

Ideally, for multiproduct equipment, a cleaning validation study would be performed for all the products manufactured in any manufacturing sequence. This undertaking is obviously beyond the resources of any company. In order to reduce the analytical workload involved in testing all the permutations of the manufacturing sequences, products and equipment are grouped into families and the worst case is selected in each family. The matrix and worst-case approaches are invaluable in establishing a manageable cleaning validation program. The matrix approach enables the use of a representative product or equipment out of all products or equipment grouped according to similarities. With the worst-case approach, on the other hand, the most stringent acceptance criteria are set to fit all the products manufactured in any sequence.

Cleaning agents are treated as special cases of the active ingredients.

Based on these approaches, a cleaning validation policy has been developed to select a product, which can represent all other products manufactured in a piece of equipment, using the same cleaning procedure. The selection of an active substance as representative of all these products is based on different characteristics of the active substances, such as: structural similarity, similar sorption/desorption isotherm behavior, similar formulation, similar potency, solubility, and degree of cleaning difficulty (Mendenhall 1989). For each characteristic, the worst case is selected out of all the products processed in a piece of equipment. These worst case characteristics are gathered to create an all-worst-case virtual product to be used as a reference product to demonstrate the effectiveness of the cleaning procedure. A similar approach for the design of the cleaning validation program makes the use of a "guiding substance" (Zeller 1993) or a "targeted substance" (Laban et al. 1997) selected to best fit predetermined criteria, such as pharmacological activity, solubility, concentration, and analytical detection.

The solubility of the active ingredient of the contaminant product in water, or any other solvent used for equipment cleaning, is a critical factor for the ease of cleaning. The more insoluble the active ingredient, the more difficult it is to get rid of it. The worst case is represented by the product with the most insoluble active ingredient and the cleaning validation study will be performed after cleaning the manufacturing equipment from this product, using an analytical method of an adequate sensitivity, developed and validated for this active ingredient.

Based on the aforementioned approach, the following policy is introduced:

> Only one product out of a group of products processed in a piece of equipment is selected for the cleaning validation study, based on the lowest solubility of the active ingredient.

Product Grouping

As for equipment, products may be grouped according to similarities: same active ingredient with several strengths or formulations differing in the inactive ingredients.

> The product worst cases are presented by the highest strength of the active ingredient and by the most difficult to clean formulation.

SELECTING THE WORST CASE RELATED TO THE EQUIPMENT

The matrix and worst-case approaches are applied in order to limit the number of pieces of equipment that have to be validated for cleaning. Taking into consideration that the product contact surfaces are mostly made of stainless steel and that similarities exist in the equipment design, operating principle and size, and in the cleaning procedure, a rationale for grouping pieces of equipment and selecting one representative piece for the cleaning validation study is developed.

Also to be addressed is the critical issue of identifying, sampling, and testing the hard-to-clean locations in the equipment.

Equipment Grouping

To make the cleaning validation program manageable, the manufacturing equipment is grouped, according to the following criteria listed, so that one piece of equipment in a group may represent the entire group. Consequently, the cleaning validation study is performed on this representative piece of equipment. The criteria for equipment grouping are listed as follows.

> Identical, interchangeable pieces of equipment with the same cleaning procedure can be grouped together.

Batch processing equipment of different sizes, where only one batch size is processed, may not be grouped as opposed to continuous processing equipment. Usually, a product is manufactured in one type and size of granulator or blender, however it can be milled, compressed, or filled in more than one type of equipment. In the latter cases, pieces of equipment with the same operating principle may be grouped.

> Equipment with the same operating principle and the same cleaning procedure, but with different product contact surface area, can be grouped, if they can be interchanged.

Since a larger batch is processed in larger equipment, the equipment with the larger surface area represents the worst case.

Hard-to-Clean Locations

The issue of the homogeneity of dispersion of residues has been discussed and the potential for pockets of residual products in poorly accessible parts of certain types of processing equipment has been stressed (Mendenhall, 1989). The "best case scenario" described by Agalloco in his theory of dispersion would be represented by freely soluble residues—as in solution formulations—as opposed to the "worst case scenario" represented by a low-level residue in solid dosage forms. Since the homogenous dispersion of the contamination in the equipment can only be assumed, the identification of the hard-to-clean locations is crucial to the cleaning validation study. The regulatory requirements are clear, and there is a consensus in the industry in this regard.

Some experiments have been made to determine the hard-to-clean locations by using water soluble (sodium chloride) or insoluble (sodium benzoate) materials or a soluble dye (sodium fluorescein) and detecting the locations that remain unclean visually or with the help of an ultraviolet (UV) light. These experiments may be valuable for small equipment or for cleaning verification of drug containers, such as glass ampoules and vials. This is not recommended, however, for larger equipment or for equipment with inaccessible parts. In these cases, the risk exists that the indicator materials may not be completely washed out on the first cleaning and will reappear in the next product manufactured. The knowledge of the equipment design and construction is of better value to assess these hard-to-clean locations.

Special consideration should be given to cleaning-in-place systems. A thorough verification of the equipment cleanliness should be performed after a CIP cycle and should include impellers, blades, screens, gaskets, valves, and instrument connections.

The worst case in relation to equipment refers to the hard-to-clean locations, such as valves and gaskets. These locations should be identified in the selected piece of equipment based on the intimate knowledge of the manufacturing equipment and its operating principle, on scrupulous visual inspection, and on cleaning experience.

Table 5.1. shows hard-to-clean locations for the most commonly used equipment for solid dosage forms. The list is indicative and not intended to be exhaustive. Moreover, since specific adaptations or modifications may have been made to the original equipment, it is advisable to carefully examine the equipment after processing, and especially after products known to be difficult to clean. Antacids are such an example, due to the large surface area of the active ingredients present in high concentration.

The worst case for a group of equipment is represented by the equipment with the larger product contact surface and the hardest-to-clean locations.

SELECTING THE SCIENTIFIC BASIS FOR THE CONTAMINATION LIMIT

Your answer to the question, How clean is clean? forms the basis of your cleaning validation program and must be scientifically justifiable. At present, there is no universally

Table 5.1. Potential Hard-to-Clean Locations in Manufacturing Equipment

Granulation		
High-shear mixers	Diosna	inner side of lid lid gasket inner side of discharge valve bottom of impeller impeller blades chopper blades vent filter
Dry Granulators	Compactor	feed hopper dosing spiral rollers oscillator blades screen
Planetary mixers		
Drying		
Fluid bed dryers	Glatt	trolley window upper cylinder wall trolley gasket filter bags
Tray ovens	All	tray corners exhaust air duct or grill
Particle Size Reduction		
Cutting mills	Fitzpatrick	cutting blades screen
Screening mills	Oscillating (Frewitt)	oscillator blades screen
	Rotating (Glatt)	rotor blades screen
Blenders		
Diffusion mixers	V-blender	inner side of lid lid gasket inner side of discharge valve intensification bar
Bins	All	inner side of lid inner side of discharge valve

Continued on next page

Continued from previous page

Unit Dosing		
Tablet press	All	feed hopper force feeder punch and die
Capsule filling machine	All	feed hopper
	Volume fill	piston
	Auger fill	plate filler
Powder filler	Auger fill	feed hopper dosing spiral dosing funnel
Coating		
Pan	All	pan exhaust air duct dosing pump baffles
Fluidized bed	Glatt	trolley window upper cylinder wall trolley gasket filter bags dosing pump

Note: The equipment name and classification are listed as given in SUPAC IR/MR–Manufacturing Equipment Addendum (FDA, January 1999).

accepted contamination limit for the active ingredient or for the cleaning agent residues. The regulatory authorities offer little help in this matter. "The firm's rationale for the residue limits established should be *logical*, based on the manufacturer's knowledge of the materials involved and be *practical, achievable* and *verifiable*. It is important to define the sensitivity of the analytical methods in order to set *reasonable limits*. Some limits that have been mentioned by industry representatives in the literature or in presentations include analytical detection levels such as *10 ppm*, biological activity levels such as *1/1000 of the normal therapeutic dose* and *organoleptic levels such as no visible residue*".

> Except for penicillin (21 CFR 211.176, Penicillin Contamination), FDA has not established standard acceptance limits for cleaning validation. Due to the wide variation in both equipment and products produced, it would be unrealistic for the agency to determine a specific limit . . . Firms need to establish limits that reflect the practical capability of their cleaning processes, as well as the specificity of the analytical test method . . . This application of the [ICH and USP] 0.1% impurity threshold is inappropriate because the limit is intended for qualifying impurities that are associated with the manufacturing process or related compounds and not extraneous impurities caused by cross contamination . . . When determining the acceptance limit, relevant factors generally include (1) evaluation of the therapeutic dose carryover; (2) toxicity of the potential contaminant; (3) concentration of the contaminant in the rinses; (4) limit of detection of the analytical test method; and (5) visual examination. While we suggest that these factors be considered, relying only on visual examination would not be scientifically sound (Human Drug cGMP Notes 1998).

It is, therefore, up to the company to set limits. The key success factors are that residue limits should be practical, achievable, and verifiable. A review of the articles, published in the recent years by industry experts and of presentations made by FDA representatives in various meetings and conferences, shows a consensus toward the adoption of a limit based on the pharmacological activity (potency) or toxicological (safety) data of the active ingredient. However, contamination limits based on other considerations have been proposed and are discussed later in this chapter.

In most cases, an arbitrary safety factor has been added in order to achieve a lower, yet reasonable, contamination level. Factors of 1/10, 1/100, and 1/1000 have been used, for benign, potent, and highly potent products, respectively (Mendenhall 1998; Flickinger 1997). The 1/1000 factor has been interpreted as follows:

> There are 3 factors of 10 in the 1/1000 fraction. The first is that pharmaceuticals are often considered to be non active at 1/10 of their normally prescribed dosage; the second is a safety factor [for the first one]; and the third is that cleaning validation should be robust, i.e. be vigorous enough that it would be considered acceptable for quite

some time in a world with ever-tightening standards (Fourman and Mullen 1993).

Another explanation to the use of a safety factor stems from the realization that, although most contamination limit determination strategies assume a uniform distribution of the contaminant, the actual distribution may not be as uniform as one would like it to be (Jenkins and Vanderwielen 1994).

The determination of the acceptable contamination limit should first take into consideration the specific manufacturing operation. For example, the following scenarios in an API facility may be different from one process to the other and they depend on the end use of the product (Lazar 1997).

- Sequential step within one process

- Nonsequential intermediate steps of one process

- Intermediate step of a different process

- Final bulk to final bulk (different products)

- Equipment dedicated to one process step

For approved pharmaceutical dosage forms or for APIs, it is usually a difficult task to set acceptance limits. For investigational drugs, this mission can be even harder due to the limited knowledge of parameters, such as therapeutic dose, toxicity, or safety. In addition, due to the versatile use of equipment at this stage of development and the fact that these drugs are subject to frequent changes in processing, formulation, and dosage strength, cleaning verification after each run is recommended. This involves testing of residuals to confirm that they have been removed to an established level, generally using an absolute limit expressed in part per million (ppm).

Contamination Limit Based on Toxicological Data

Long-term toxicity data for humans are lacking for most chemical compounds and therefore acute and chronic toxicity studies on laboratory animals of the mammalian species are evaluated to estimate potential toxic effects on man (Layton et al. 1987). Toxicological assessments normally include acute toxicity data, using different routes of administration as well as mutagenicity, carcinogenicity, teratogenicity, sensitization, and irritation. For the purpose of assessing human toxicological effects, various indices have been used, such as the no-observed-effect level (NOEL), the lethal dose (LD50), and the acceptable daily intake (ADI).

- The NOEL is usually based on a two-year chronic toxicity study on various mammalian species and is defined as the highest dose at which no toxic effects are observed (Layton et al. 1987). OSHA (Occupational and Safety Health Organization, USA) developed this index to limit worker exposure to various chemical compounds in the work place (Agalloco 1992).

- The LD50 is based on an acute toxicity study on various mammalian species and is an estimate of the lethal dose to 50% of the animals, following a single dose (Layton et al. 1987). This value is mostly used for nonactive ingredients such as cleaning agents or sanitizers (Agalloco 1992).

- ADI is defined as the amount of toxicant in milligrams per kilogram body weight per day which is not anticipated to result any adverse effects after chronic exposure to the general population of humans (Dourson and Stara 1983). The ADI is difficult to establish and can be estimated based on the LD50 using a factor of 5×10^{-6} to 1×10^{-5} that incorporates variability associated with animal-to-human toxic response and dose equivalence (Layton et al. 1987).

ADIs are usually associated with food additives, vitamin preparations, and other ingested non-pharmaceutical products. The acceptable level would depend on whether a drug is administered to treat an ailment, or whether it was carried through a manufacturing process as an impurity. The acceptable daily intake level of a chemical administered to relieve an illness would be essentially identical to the lowest therapeutic dose. The acceptable level of the chemical present as an impurity would be no higher than the NOEL.

NOELs are occasionally included in computer accessed safety databases. However, the term NOEL is used to describe the highest dose that did not cause a measurable or observable effect in a specific test. The NOEL for an acute oral toxicity test will be different than the NOEL determined for teratology, neurotoxicity, subchronic toxicity, acute inhalation, and other tests. The NOEL depends on the route of exposure (oral, dermal, inhalation, etc.), duration of exposure (single dose or multiple doses), species in which the test was conducted, and the endpoint measured.

Another approach makes use of the compound's LD50 to calculate the NOEL, which is used with an animal-to-human uncertainty factor and the average adult weight to derive the ADI (Layton et al. 1987; Conine et al. 1992; Kirsch 1998).

The previously mentioned toxicological indices are converted in conjunction with the dosage regimen and the batch size to the maximum allowable carryover to the next product. This holds true for APIs as well: "[The] amount of residue [is] assumed to be transferred to the finished dosage product at its highest dosage level. The theoretical amount of contamination should be calculated for one dosage unit and compared with the set limit".

In general, it is logical to set residual limits on a pharmacologically defined concentration based on the potency of the substance. However, with less potent drugs, this could lead to intolerably high residual limits (Zeller 1993).

Dose-Related Contamination Limit

This limit is based on the lowest therapeutic dose (LTD). The LTD is defined as the lowest dose in any of the active ingredients that may be given to a patient in any dosage form available, and is generally equivalent to one unit dose, i.e., one tablet or one capsule.

Contamination Limit Based on Other Considerations

- Not Detectable

  This type of limit refers to a specific analytical method. The disadvantage of this type of limit lies in the fact that the detection level decreases as the sensitivity of new analytical methodology increases. The setting of the contamination limit based on the detection level of the analytical method requires a full revalidation of the cleaning procedure with the introduction of the more sensitive method and therefore should be avoided.

- Absolute Limit

  An absolute limit also termed "single limit" or "blanket specification" means that the same limit is set for any product, without consideration to toxicological data or to the detection level of the analytical method, and is usually expressed in parts per million (ppm) (McCormick and Cullen 1993). In this approach, it is suggested to set limits in line with residual limits set for hazardous compounds, such as pesticides in food (Harder 1984). This recommendation is based on 21 CFR, Part 193 "Tolerances for Pesticides in Food Administered by the Environmental Protection Agency". Based on the fact that the amount of administered drug is much lower than the quantity of food that would be eaten, limits in the ppm range—acceptable for hazardous compounds in food—seem to be reasonable. This type of limit is also based on an acceptable contamination level set by official pharmacopeias, such as the heavy metals or arsenic tests. A limit of not more than 10 ppm seems to be widely used (Fourman and Mullen 1993), although lower limits, as low as a few ppm or even less, have been used by some firms (McCormick and Cullen 1993; Kirsch 1998).

Deciding upon an allowable contamination level irrespective of the potency of the drug reduces the quality of the product or the cost effectiveness of manufacturing. For drugs with low pharmacological potency, this decision results in overcleaning while for high potent drugs it leads to insufficient cleaning (Zeller 1993).

A variant of the absolute limit, which is seldom used, requires the contamination to be less than 1 mg/m^2 of the product contact surface area (Laban et al. 1997).

- Combination of Absolute Limit and Pharmacological Limit

  An approach combining the absolute limit and the pharmacological limit was suggested (Zeller 1993), so that the allowable contamination level has to decrease with minimum therapeutic doses in a linear relation for all therapeutic doses lower than 1 mg.

In the case of products for external use, the limit could be increased by an order of magnitude, due to their low bioavailability, when compared to that of products administered orally. However, for parenterals, the acceptable contamination level could be halved in comparison to oral drugs (Zeller 1993).

- Log Reduction of Contaminant Concentration
  This type of acceptance criterion is based on the reduction by an arbitrary number of orders of magnitude, usually 3 logs (one log is a tenfold reduction), in the concentration of the contaminant, measured before and after cleaning of the equipment. This type of criterion is widely used in the validation of aseptic processes, for example, to determine the effectiveness of the depyrogenation process. Without correlation to adverse effects of the contamination in humans, this type of limit cannot be meaningful.

Acceptance Criterion Based on Visual Inspection

This acceptance criterion states that no residues may be visible in the equipment after cleaning and drying. While this criterion may sound too unsophisticated and non-quantitative, it was stressed that "the visual cleanliness criterion was more rigid and clearly adequate" in comparison to quantitative calculations. Therefore, the visual inspection must be part of the cleaning procedure, even though the cleanliness level of the equipment is chemically determined. It has also been shown that, while other criteria such as LTD/1000 or 10 ppm were met, residues could still be visible after cleaning of equipment in which sodium chloride tablets were processed (Fourman and Mullen 1993). The visual detection limit of most active ingredients, according to the same source, is approximately 100 μg per 25 sq. cm. Clearly, for high potent or toxic drugs, the calculated allowable amount of residuals may be well below visual detectability.

The use of a set of criteria is frequently implemented. The approach of setting few criteria is quite known. This approach is presented already by Fourman and Mullen (1993) using dose-related limit together with 10 ppm absolute value and visual inspection. The same principle is presented by Forsyth and Haynes (1999) using safety (ADI) as the first criterion. The second criterion is an adulteration limit which states that no more than 10 ppm of any product will appear in another product. The third criterion is that the equipment must be visually clean.

The Proposed Approach

The approach proposed here for setting the residue limit for the active ingredient uses the LTD, as the most easily available index and the most widely used in the industry and for the cleaning agent the LD50, both with the addition of an arbitrary safety factor of 1/1000.

The acceptance criteria of cleaning validation may be formulated as follows:

> The maximum contamination level of active ingredient from any product in the maximum daily dose of another product is not more than 1/1000 of the LTD of the former product.

and

The maximum contamination level of cleaning agent in the maximum daily dose of any product is not more than 1/1000 of the LD50 of the cleaning agent.

---

and

No visible residue must be found in the equipment, after cleaning.

SELECTING THE SAMPLING METHOD

There are three known sampling methods: the swabbing (or direct surface sampling) method, the rinse sampling method and the placebo method. A description of the methods, their advantages, and their disadvantages are presented from regulatory as well as practical viewpoints.

The Swabbing Method

This method is based on the physical removal of residue left over on a piece of equipment after it has been cleaned and dried. A swab wetted with a solvent is rubbed over a previously determined sample surface area to remove any potential residue, and thereafter extracted into a known volume of solvent in which the contaminant active ingredient residue is soluble. The amount of contaminant per swab is then determined by an analytical method of adequate sensitivity.

> The advantages of direct sampling are that areas hardest to clean and which are reasonably accessible can be evaluated, leading to establishing a level of contamination or residue per given surface area. Additionally, residues that are 'dried out' or are insoluble can be sampled by physical removal (FDA, July 1993).

Advantages and Disadvantages

The following are the major advantages of the swabbing method.

- Direct evaluation of surface contamination.

- Insoluble or poorly soluble substances may be physically removed from the equipment surfaces.

- Hard-to-clean but accessible areas are easily incorporated into the final evaluation.

The main disadvantages of the swabbing method are these:

- Difficult to implement in large-scale manufacturing equipment.

- Extrapolation of results obtained for a small sample surface area to the whole product contact surface of the equipment.

- Less accessible or hard to clean areas are more difficult to sample and may require a combination of the swabbing method with the rinse sample method.

Product may accumulate in these hard-to-clean locations leading to a high amount of active ingredient in a small surface area. Therefore, for the results to be significant, the contribution of the hard-to-clean location surface areas to the whole equipment product contact surface area must be taken into account in the calculation of the contamination limit (Lazar 1997).

Although swabbing is not the preferred method for routine testing of large-scale manufacturing equipment, the direct evaluation of surface contamination makes it a most valuable tool in a cleaning validation study (Smith 1992).

Swab Characterization

Natural or synthetic cotton wool, glass wool, synthetic fabric, and filter paper are commonly used as the carrier for removing the contaminant from the equipment surface. The selection of the best swab material is guided by physical as well as chemical criteria, with the objective of having—with no previous treatment—good physical removal properties and the least significant chemical background contribution. The selection of the swab material is based on the following criteria.

- Swab absorption capacity:
  The swab material must have a solvent absorption capacity sufficient to be moistened or saturated in order to add a solubilization effect to the physical removal process.

- Swab interference:
  Particles, fibers, and lints released in swab solvent extract obscure the light beam path in spectrophotometric methods and should be filtered out before testing. Lint-free, long fiber fabrics or wool, and quantitative filter paper are the preferred materials. Swab extracts are tested using solvents which will normally be used in sampling and analytical testing: water, ethanol, diluted acids and alkalis.

Substances extracted by solvents from the swab must not interfere with the analytical detection method. As with the previous property, this one is also critical and must be checked in the characterization study to be performed for the analytical detection method. When testing swab samples, a blank determination of a virgin swab solvent extract is mandatory to eliminate extraneous factors that may influence the quantitative determination of the contaminant.

Some swabbing materials, such as quartz wool and synthetic fabrics, have been screened in connection with the use of Total Organic Carbon (TOC) as the analytical detection method (Jenkins et al. 1996). The use of low-carbon leaching swabs is recommended when TOC is the analytical detection method (Lombardo 1995). A filter paper swab has been used with Fourier Transform Infra Red (FTIR) (LeBlanc 1993).

Swab Recovery

A swab recovery study must be performed to determine the ability of the swab to quantitatively remove the contaminant from the surface sampled.

The roughness and the surface area of the swab material are also of importance in the physical removal process. Swab materials that are too smooth or too coarse do not ensure complete removal of the contaminant. Synthetic lint-free fabrics or synthetic cotton wool in which the fiber length is controlled to a certain extent and quantitative filter paper, such as Whatman #42 or equivalent, are materials of choice.

Another parameter which might influence the recovery is the ability of the swab to absorb or adsorb the contaminant. This is a critical property which must be checked in the recovery study by spiking swabs with known concentrations of the contaminant and then extracting it.

The recovery factor must be taken into consideration in the calculation of the allowable contamination limit. This is particularly important due to the fact that the result obtained from a small surface area is extrapolated to the entire equipment product contact surface. A recovery factor of 70% is acceptable, however, factors as low as 50% may be obtained. In cases where low results are obtained in a reproducible manner, the sample surface area may be sampled again using a second swab and the results obtained from both swabs added together.

The swab recovery study is generally performed on pieces (also called coupons) of the same construction material as the equipment. An area equivalent to the equipment sample surface area is swabbed and tested.

The swab recovery study may be performed as part of the validation of the analytical method.

Sampling Locations and Number of Samples

The sampling locations are dictated by worst-case conditions. The equipment's hard-to-clean locations are identified, based on cleaning experience and the intimate knowledge of the equipment design, including the engineering piping and instrumentation diagrams (P&ID). Before starting the cleaning validation study, it may be useful to perform a contamination mapping of the equipment, including the expected hard-to-clean locations, by simulating the contamination with soluble and insoluble materials (such as sodium chloride and sodium benzoate, respectively), using rapid analytical methods of moderate specificity.

Since the homogeneity of the contaminant on the equipment product contact surface can only be assumed, several samples, but not less than three samples per piece of equipment, must be taken, including the hardest-to-clean locations. The number of samples should take into consideration the equipment surface area, design, shape, operating principle, and construction materials.

Sample Surface Area

Sample surface areas usually vary from 25 sq. cm to 100 sq. cm and should be large enough to allow the recovery of the contaminant in a quantity sufficient to be detected by the analytical method. This small sample surface area is assumed, by extrapolation, to represent the amount of residual contaminant in the whole equipment surface area. Stainless steel templates are used on flat accessible surfaces to delimit the area to be swabbed. For more reproducible results in hard-to-clean locations, it is recommended, where feasible, to swab the whole area.

Swabbing Technique

As a manual and mechanical operation, the swabbing method is difficult to standardize. Operators performing the sampling must be adequately trained and given precise instructions in order to obtain reproducible results. The full coverage of the sample surface area, the amount of pressure to be applied, the amount of solvent absorbed by the swab and the strokes' direction must be emphasized in the operators' training. The swabbing technique must describe, in detail, the direction and numbers of strokes (or passages) needed for the full coverage of the sample area in a reproducible manner. Such techniques have been described in the literature (Lombardo 1995; Shea et al. 1996).

Swab Saturation and Extraction Solvents

The solvents used for swab saturation and extraction may be one and the same. The selection of the swab saturation solvent must take into consideration the solubility of the contaminant active ingredient, so that the recovery is the highest and reproducible. The toxicity of the solvent should also be considered to protect the operator and to prevent solvent residues from being transferred to the next product. The most useful solvent is ethanol, which can readily dissolve a large number of compounds, evaporates quite rapidly, and has a low toxicity compared to other solvents.

The nature of the swab extraction solvent is dictated by the analytical method, taking into consideration its dissolution properties as well as the analytical technique used.

The Rinse Sample Method

This method is based on the analytical determination of a sample of the last rinsing solvent (generally water) used in the cleaning procedure (LeBlanc 1998). The volume of solvent used for the last rinse must be known to allow for the quantitative determination of the contaminant. The rinsing sample method is part of the cleaning procedure and therefore does not require any special instruction to be performed.

Two advantages of using rinse samples are that larger surface areas may be sampled and inaccessible systems or ones that cannot be routinely disassembled can be sampled and evaluated. A disadvantage of rinse sample is that the residue or contaminant may not be soluble or may be physically occluded in the equipment (FDA, July 1993).

Advantages and Disadvantages

The following are the major advantages of the rinse sample method.

- Ease of sampling.

- Evaluation of the entire product contact surface.

- Accessibility of all equipment parts to the rinsing solvent.

- Best fitted to sealed or large-scale equipment and equipment which is not easily or routinely disassembled.

The main disadvantages of the rinse sample method are these:

- No physical removal of the contaminant.

- The rinsing solvent may not reach inaccessible or occluded equipment parts.

- Use of organic solvents for water insoluble materials.

Although the rinsing sample method has its limitations, it is widely used in large-scale equipment, especially in the API industry, where equipment is cleaned by rinsing and refluxing with the appropriate solvent.

The Placebo Method

The technique known as the placebo method is seldom used. This method is based on sampling and testing of a portion of a placebo batch processed in a piece of equipment after cleaning and drying. The regulatory authorities vigorously discourage the use of this method. The FDA has identified disadvantages to the placebo method.

One cannot ensure that the contaminant will be uniformly distributed throughout the system. . . . Additionally, if the contaminant or residue is of a larger particle size, it may not be uniformly dispersed in the placebo. Some firms have made the assumptions that a residual contaminant would be worn off the equipment surface uniformly; this is also an invalid conclusion. Finally, the analytical power may be greatly reduced by dilution of the contaminant. Because of such problems, rinse and/or swab samples should be used in conjunction with the placebo method (FDA, July 1993).

The preferred sampling method and the one considered as the most acceptable by regulatory authorities is the swabbing method.

SELECTING THE ANALYTICAL METHOD

Basic Requirements

A wide variety of sensitive analytical methods is presently available (Smith 1993; Gavlick et al. 1995). However, the detection of residues for cleaning validation purposes present a special challenge to the analytical scientist. The regulatory authorities put an emphasis on the development of specific and sensitive methods.

> Determine the specificity and sensitivity of the analytical method used to detect residuals of contaminant. With advances in analytical technology, residues from the manufacturing and cleaning processes can be detected at very low levels. If levels of contamination or residual are not detected, it does not mean that there is no contaminant present after cleaning. It only means that levels of contaminant greater than the sensitivity or detection limit of the analytical method are not present in the sample. The firm should challenge the analytical method in combination with the sampling method(s) used to show that contaminants can be recovered from the equipment surface and at what level, i.e., 50% recovery, 90%, etc. This is necessary before any conclusions can be made based on the sample results. A negative test may also be the result of poor sampling technique (FDA, May 1993).

The very low contamination limit, resulting from the use of high safety factors and worst cases, as well as the analytical methods available, impose contingencies in the analytical method development (see Table 5.2).

Table 5.2. The Basic Requirement for the Analytical Method

---

1. The sensitivity of the method shall be appropriate to the calculated contamination limit.

2. The method shall be practical and rapid, and, as much as possible use instrumentation existing in the company.

3. The method shall be validated in accordance with ICH, USP and EP requirements.

4. The analytical development shall include a recovery study to challenge the sampling and testing methods.

Analytical Method Survey

Specific Methods

- Chromatographic methods such as LC/MS, GC/MS, and HPLC

  Advantages:
  These chromatographic methods are the methods of choice, as they separate analytes, are highly specific, highly sensitive, and quantitative. Various detection techniques may be used to enhance these properties and to allow for the determination of a wide variety of active pharmaceutical ingredients and cleaning agents. Among these techniques, the following are commonly used: spectrophotometric, electrochemical, fluorescence, and refractive index (Kirsch 1998; Mirza et al. 1998).

  Disadvantages:
  These methods are costly and time-consuming.

- Thin layer chromatography (TLC)

  Advantages:
  TLC is highly specific and moderately sensitive. Various detection techniques may be used to enhance the visualization and determination of the analytes: UV and visible wavelength and colorimetry.

  Disadvantages:
  The method is semiquantitative and therefore may be used only to give an indication of the equipment cleanliness level at the preliminary stage of development of the cleaning procedure.

- Specific ion meter

  Advantages:
  The quantitative determination of an active ingredient is rapid and specific.

  Disadvantages:
  The method can only be used if the active ingredient alone possess a specific ion.

Nonspecific Methods

Although not recommended, nonspecific, yet rapid methods, have been used, sometimes with modification in order to achieve a meaningful specificity.

- Spectrophotometric methods in the visible, infrared, or UV ranges

  Advantages:
  These methods are rapid, inexpensive, and moderately specific. Fourier transform infrared spectroscopy has been used as a primary qualitative tool to obtain a finger print of cleaning agent residues before their quantitative analysis (LeBlanc 1993).

Disadvantages:
Active ingredients can only be quantitated, if they possess a strong chromophore that can be detected at very low levels and when the inactive ingredients do not interfere with active ingredient at the same wavelength. In some cases where the absorption of the inactive ingredients is low, correction may be made by subtracting the absorbance of the formulation placebo.

- Total organic carbon (TOC)

Advantages:
The TOC method offers, at a moderate cost and in addition to its rapidity, a detection capability down to the ppb range. The method can also be applied to on-line measurements. TOC in conjunction with other methods, such as conductivity and pH, has been used to demonstrate the absence of residual cleaning agent in rinse samples (Jenkins et al. 1996; Holmes and Vanderwielen 1997).

Disadvantages:
The limitation of the method resides in the fact that only water-soluble compounds can be analyzed and that most, if not all, inactive ingredients are organic compounds which contain carbon atoms and thus interfere in the measurement of the active ingredient.

- Other Methods
Nonspecific methods such as pH, conductivity, and total solids have very limited use and can only be used in conjunction with a specific method.

In conclusion:

> The chromatographic methods are preferred for cleaning validation studies because of their sensitivity, specificity, and ability to quantify.

Analytical Method Validation

The validation of the analytical method follows ICH guidelines (1995, 1997), USP (1995). In addition to these basic requirements, all the cleaning validation elements related to the sampling method, such as swab interference, placebo, solvent, shall be verified through the validation of the analytical method.

The method validation should demonstrate:

- Accuracy (recovery) and range

- Precision

- Linearity

- Selectivity

- Detection Limit

- Quantitation Limit

- Stability of standard and sample solutions

Analytical Method Implementation

The analytical method, once validated, is transferred to the quality control (QC) laboratory for implementation and routine testing. The reproducibility of the method between the analytical development laboratory and the QC laboratory is part of the technology transfer and of the method validation.

6

Contamination Acceptance Limits

The basic elements of a cleaning validation program relate to the products, the equipment, the cleaning procedures and the sampling method. The characteristics, which must be taken into account for each element, include (see Table 6.1):

- Characteristics related to the contaminating product, such as its therapeutic dose and the solubility of its active ingredient, as it relates to cleanability

- Characteristics related to the contaminated product, the next produced product, such as the batch size and the maximal daily dose (the number of tablets or capsules taken in one day)

- Characteristics related to the equipment, such as the product contact surfaces and the hard-to-clean locations

- Characteristics related to the cleaning procedure, such as the cleaning agents

- Characteristics related to the sampling method, such as the sample surface area and the recovery

Let us develop together the matrix and worst-case approaches. Assume that you are manufacturing only two products, using only one piece of equipment. The two possible manufacturing sequences are: Product A before Product B or Product A after Product B. For equipment cleaning validation in this simplified case, your goal is to demonstrate that not more than an acceptable amount of active ingredient residue of Product A (or Product B) can be detected in the maximal daily dose of Product B (or Product A). The characteristics of Product A and Product B are tabulated in Table 6.2.

Each of the product characteristics has a different effect on the calculation of the allowable amount of contaminant carried over to the next product and this amount has to be the lowest to present the worst case scenario. It follows that the product with the lowest therapeutic dose should be selected so that the minimal amount of contaminant will be left after cleaning. The product with the highest daily dose should be selected so that when the batch is divided into the maximum number of units, each unit will contain the minimal residual amount of contaminant. The product with the smallest batch size should be selected so that the overall quantity of residues allowed for each manufactured batch will be minimal. To illustrate the concept, let us go back to our

Table 6.1. Cleaning Validation Elements

Element \ Characteristic	Therapeutic Dose Activity/Toxicity	Solubility	Batch Size	Maximal Daily Intake	Product Contact Surfaces	Hard-to-clean Locations	Sample Area	Recovery
Contaminant product	▓	▓						
Contaminated product			▓	▓				
Equipment					▓	▓		
Sampling method							▓	▓
Cleaning agent	▓							

Table 6.2. Product Characteristics

	Therapeutic Dose	Maximal Daily Dose	Batch Size
Product A	low	low	small
Product B	high	high	large

previously mentioned Product A and Product B scenario and determine the worst case values to be selected (see Table 6.3).

We have thus created a "virtual" product, which, for the sake of cleaning validation, combines all the worst case parameters. In practice, when many products are manufactured in the same piece of equipment, the lowest therapeutic dose, the maximal daily dose, and the smallest batch size of different products will be used to calculate the acceptance limit.

> The contamination acceptance limit is calculated using the virtual product worst case characteristics: lowest therapeutic dose, maximal daily dose, and smallest batch size.

LOWEST THERAPEUTIC DOSE (LTD)

The product with the lowest active ingredient therapeutic dose (LTD) represents the worst case and this value is used to calculate the acceptance criterion of the cleaning validation study for the group of products manufactured in one equipment group. The LTD, as defined in Chapter 5, may be found in national or international medical compendia, such as the Physician's Desk Reference (USA current edition), Martindale–The Extra Pharmacopoeia (U.K., current edition), Vidal (France, current edition), Rote Liste (Germany, current edition).

By using a safety factor of 1/1000, LTD/1000 represents the maximum amount of contaminant that may be carried over to the maximal daily dose of the next product.

Table 6.3. Product Characteristics—Worst Case Scenario

	Therapeutic Dose	Maximal Daily Dose	Batch Size
Product A	low	low	small
Product B	high	high	large
Selected Product	A	B	A

MAXIMAL DAILY DOSE

The maximal daily dose (D), also named the maximal daily intake, is the maximal dose in mg that should not be exceeded. This data also appears in the aforementioned compendia.

The contaminant product residues are assumed to be spread all over the equipment product contact surfaces and to be divided equally in all the units (tablets or capsules) of any other product batch subsequently processed in the same piece of equipment. The higher the maximal daily doses of the subsequent product the bigger the amount of contaminant that will be absorbed in one day by the patient. The worst case is therefore represented by the product with the highest maximal daily dose. The above mentioned assumption seems contradictory to the FDA statement "One cannot ensure that the contaminant will be uniformly distributed through the system" (FDA, July 1993). However, as already discussed, the sampling and testing of the equipment hard-to-clean locations ensure that accumulation of contaminant in such a location is taken into account.

$\dfrac{LTD/1000}{D}$ represents the maximum allowed amount of contaminant in the maximal daily dose of the contaminated product.

HIGHEST UNIT DOSE WEIGHT AND SMALLEST BATCH SIZE

Furthermore, the same amount of contaminant may be divided in a high (for example, 1,000,000 tablets) or in a small (for example, 100,000 tablets) number of units of the subsequent product. For two different products with the same maximal daily dose, the maximal daily dose of the contaminated product with the smaller batch size will contain a higher amount of contaminant. The worst case here will therefore be the smallest number of units in a batch of product manufactured in the same piece of equipment. For the virtual all-worst-case product, the smallest number of units is obtained by dividing the smallest batch size in weight (Wb) for any product by the highest unit dose weight (Wt) for any product in the product group (see Table 6.4).

Table 6.4. The Worst Cases as They Relate to Products

1. Lowest therapeutic dose

2. Highest maximal daily dose

3. Smallest batch size

4. Highest unit dose weight

CALCULATION OF THE SAMPLE CONTAMINATION LIMIT FOR THE ACTIVE INGREDIENT

From the above principles and the selected characteristics for the hypothetical all-worst-case product, the following equation has been developed to calculate the maximum allowable amount of contaminant (MC) in mg per swab, for a specific piece of equipment:

$$MC \ (mg/swab) = \frac{LTD/1000}{D} \times \frac{Wb}{Wt} \times \frac{Ss}{Se} \times R$$

Where:

LTD	– Lowest Therapeutic Dose	(mg)
D	– Highest Maximal Daily Dose	(Dose units)
Wb	– Smallest Batch Size	(g)
Wt	– Highest Unit Dose Weight	(g)
Ss	– Swab Area	(cm²)
Se	– Equipment Product Contact Surface Area	(cm²)
R	– Recovery Factor of Active Ingredient with Lowest Solubility	(%)

From the above equation, it can be seen that the most stringent acceptance criterion, i.e., the lowest contamination level, is obtained when the values for *LTD* and *Wb* are the lowest while the values for *D* and *Wt* the highest.

CALCULATION OF THE SAMPLE CONTAMINATION LIMIT FOR THE CLEANING AGENT

The same approach can be adopted to calculate the contamination limit for the cleaning agent, using LD50. When formulated cleaning agent is used, the LD50 of the most toxic component is used. LD50 values can be found in the literature.

DISCUSSION OF THE SAMPLE CONTAMINATION LIMIT

The calculation of the sample contamination limit is the result of the development of the cleaning validation rationale, based on scientifically justified assumptions. However, there are limitations that should be kept in mind as well as practices that should be avoided.

Averaging of Sample Swab Results

Swab samples are taken from various locations using templates of a known surface area, and from the hard-to-clean locations. Equipment of complex geometry may be decomposed in portions, such as the conical or cylindrical portions, for ease of calculation of the portion surface area. The sum of the surface areas of all portions is equal to the entire surface area of the equipment. The swab samples taken from the various equipment portions are extracted and tested separately and the amount of material per swab is calculated according to the above mentioned equation. Satisfactory results are obtained when the amount of material (mg/swab) in any equipment portion is lower than the calculated sample contamination limit for this portion. Each swab result must individually comply with its specific limit. The results obtained for the individual swabs may be added taking into account their relative contribution to the contamination level. Averaging of results is not acceptable to the regulatory authorities. Out-of-specification results for any of the swabs deny the assumption of homogeneous contaminant dispersion and lead to the possibility that some of finished dosage form units may contain an amount of contaminant in excess of the allowable limit.

The measurement of the surface area of a hard-to-clean location (such as the discharge valve of a twin shell blender) is more difficult than that of large smooth surfaces and has to be measured with the best approximation. Apart from this, a hard-to-clean location is treated just like any other sampling location.

Equipment Train

Any pharmaceutical product is processed through an equipment train, e.g., for a tablet, a granulator, a mill, a blender and a tablet press are used. Each piece of equipment makes its own contribution to the amount of contaminant that may be present in the finished dosage form. Therefore the worst-case results, obtained for each piece of equipment are summed up to give the total amount of contaminant in the equipment train. The worst-case carryover of contaminant to the next product is obtained by dividing the total amount of contaminant in the train by the smallest number of units of product processed in the train.

Although this approach gives an overall estimation of the amount of contaminant in the next product, it involves additional calculation. Considering the safety factors introduced in the equation, it is deemed sufficient to use the single criterion stating that any individual swab sample result must not exceed the calculated swab sample contamination limit.

The most desirable approach is to have one set of acceptance criteria. The FDA discourages the use of variable acceptance criteria for individual swab results based on location, the averaging of individual swab results or the use of a single acceptance criterion based on cumulative residues from swab samples. The concern is again potential masking of excessive variability of an ineffective cleaning procedure and the disproportionate dilution (averaging) of high results.

The FDA expects swab sampling at multiple sites representative of the product contact equipment surface, including hard-to-clean locations. Each swab sample needs to be evaluated against a single acceptance criterion. During validation, single site sampling (i.e., worst-case) may be questioned, even if there is hard data that supports the worst case site selection. It would also be expected that, if the worst case swab samples show out-of-specification results, improved cleaning would need to be implemented to resolve this situation. However, for routine monitoring after cleaning validation, the use of the most difficult to clean location for determining the reproducibility of the cleaning procedure would be acceptable to regulatory authorities.

7

Cleaning Validation
Planning and Execution

INTRODUCTION

The next step in our validation program is to plan the activities to be performed based on the cleaning validation policy developed in Chapter 5, and then execute a cleaning validation study. A systematic approach must be developed in order to program parallel activities, which involve different groups in the organization, and to conduct the cleaning validation study in a timely and cost-effective manner.

ORGANIZATION

The program, after having been developed, must be presented and explained to the various groups in the company (R&D, production, production planning, engineering, quality assurance and quality control) taking part in the validation studies, whether their participation is active or supportive only. The close cooperation between these groups, their understanding of the fundamentals of validation and of cleaning validation in particular, and their adequate and timely contribution are all vital to the successful management of the cleaning validation program.

In most companies, the cleaning validation program is managed by the quality assurance unit, but it could be managed by any other unit depending on the company organization, provided that the program is supervised and controlled by the quality assurance unit. It is advisable to establish a dedicated cleaning validation team or unit—at least at the learning stage of the program—to provide a full overview and a focused attention to all the aspects of cleaning validation. A steering committee consisting of representatives of the aforementioned disciplines should also be instituted to deal with changes in the policy, protocol deviations, and with any unexpected event or result in the course of the validation study.

The typical responsibilities of the various groups involved in the following cleaning validation program may differ from company to company. While the repartition of the responsibilities is obvious for most of the groups, R&D, the validation group, and QA assume interchangeable roles depending on the company organization.

Validation Unit

Responsibilities of the validation unit include:

- Prepare validation master plan, working plan, and protocol, including preparation of database, grouping, determination of worst cases, and identification of hard-to-clean locations.
- Calculate the contamination limit for the active ingredient and the cleaning agent.
- Conduct validation study including sampling.
- Prepare validation report.

Production

Responsibilities of the production group include:

- Approve the validation master plan and the working plan.
- Verify accuracy of the cleaning procedure.
- Identify the equipment hard-to-clean locations.
- Perform cleaning.

Production Planning

Responsibilities of the production planning group include:

- Provide all information to build database.

Engineering

Responsibilities of the engineering group include:

- Verify accuracy of drawings and calculate product contact area.
- Assist in the identification of equipment hard-to-clean locations.

Quality Assurance

Responsibilities of the quality assurance group include:

- Approve the validation master plan and the working plan.

- Approve validation protocol.

- Oversee the validation study.

- Approve validation report.

R&D Analytical Development

Responsibilities of the R&D group include:

- Develop and validate the analytical method.

Quality Control

Responsibilities of the quality control group include:

- Perform the recovery study.

- Test samples and prepare analytical report.

PLANNING

The planning of the cleaning validation program consists of a series of steps starting from the establishment of a database and, resulting, through the grouping and the worst-case approaches of equipment and products, in the selection of the specific equipment and the representative product on which cleaning validation is performed. The sequential steps are described in Table 7.1.

In the following sections, a step-by-step description will lead the reader through the whole process of building the cleaning validation master plan. An example of a short list of equipment and products is provided to help the reader to follow the process.

Table 7.1. Planning Stages for Cleaning Validation

---

1. Database (* Equipment List, * Product List)

2. Grouping (* Equipment, * Products)

3. Worst-case selection (* Equipment, * Product)

4. Calculation of the contamination limit

5. Validation master plan

6. Working plan

7. Execution

DATABASE

The list of products and cleaning agents and the list of equipment are the basic documents on which the cleaning validation database is built. These lists must be controlled, by date and edition, verified to be accurate at the time the cleaning validation study starts and maintained up-to-date through the change control system. The lists represent the collection of the equipment and the product characteristics that will enable the grouping and the selection of the worst cases and that must be taken into account to calculate the contamination limit.

A list of equipment pieces used for manufacturing the products is prepared using the list of products and the product master batch documents. An example of such lists are shown in Tables 7.2 and 7.3.

In total, the products in Table 7.2 are manufactured using 13 pieces of equipment.

EQUIPMENT GROUPING

First, the equipment pieces are categorized according to technological groups: mixers, dryers, mills, and so forth.

From Table 7.4, it can be seen that equipment nos. 4, 7, 10, and 11 are dedicated to products E, D, C, and A, respectively. As such, they are not included in the cleaning validation program and are therefore deleted from the table.

According to the cleaning validation policy described in Chapter 5, as it relates to equipment, grouping is done by comparing the operating principle, the product contact surface area, and the cleaning procedure of the pieces of equipment. The grouping rationale must be justified and documented.

The cleaning procedure used for cleaning each piece of equipment after the manufacture of each product is verified. Usually one cleaning procedure is used for cleaning one piece of equipment. However, for certain products manufactured in the same piece of equipment, another cleaning procedure may be used, due to difficulties in removing residues from these products. In this case, a cleaning validation study has to be performed on the same piece of equipment for each cleaning procedure. It is assumed in Table 7.4 that only one cleaning procedure is used to clean one piece of equipment from all the products.

The equipment operating principle is a characteristic that differentiates pieces of equipment used for the same process: both a high-shear mixer and a fluid bed dryer may be used for wet granulation, but they are considered as different classes of equipment. A high-shear mixer and a low-shear mixer belong to the same class, but are considered as subclasses of equipment used for the same process.

The product contact surface area represents a characteristic of the equipment that is taken into account in the calculation of the contamination limit. This value is proportional to the equipment size. At this stage, the equipment size is used for grouping and is listed in Table 7.5.

Table 7.2. List of Products

Product A	Product B	Product C	Product D	Product E

Table 7.3. List of Equipment per Product

Product A	Product B	Product C	Product D	Product E
Mixer #1	Mixer #2	Mixer #2	–	Mixer #1
Dryer #1	Dryer #1	Dryer #1	–	Dryer #2
Mill #1	Mill #2	Mill #1	Mill #3	Mill #2
Blender #1	Blender #2	Blender #3	Blender #1	Blender #2
Tablet press #1	Tablet press #2	Tablet press #3	Tablet press #3	Tablet press #2

Table 7.4. Equipment Categories

Equipment No.	Equipment Name	Products
1	Mixer #1	A, E
2	Mixer #2	B, C
3	Dryer #1	A, B, C
4	Dryer #2	E
5	Mill #1	A, C
6	Mill #2	B, E
7	Mill #3	D
8	Blender #1	A, D
9	Blender #2	B, E
10	Blender #3	C
11	Tablet press #1	A
12	Tablet press #2	B, E
13	Tablet press #3	C, D

The following rationale is used to group the equipment listed in Table 7.5: Mills #1 and 2 are grouped together because they have the same operating principle. Both are screening mills with an oscillating bar and are of the same size. Blenders #1 and 2 cannot be grouped. Although both are twin shell blenders, they are different in size.

The products manufactured in a group of equipment are also grouped. The result of the grouping is shown in Table 7.6.

SELECTION OF THE EQUIPMENT WORST CASE

The contamination limit equation, developed in Chapter 6, shows that the larger the equipment product contact surface, the lower is the amount of contaminant allowed to be carried over in the next product batch. It follows that the worst-case equipment in the group, on which the validation will be conducted, is the one with the largest product contact surface.

As a result, a shorter list of equipment, on which the cleaning validation study has to be performed, is obtained. In this list, 8 out of 13 pieces of equipment—representing the 8 groups—will participate in the cleaning validation program. As the number of pieces of equipment increases and there is more interchangeable equipment, the list of equipment to be validated for cleaning is dramatically reduced. Table 7.7 shows the list of equipment on which the cleaning validation study will be performed.

Table 7.5. Equipment Grouping

Equipment No.	Equipment Name	Operating Principle	Size	Products
1	Mixer #1	High-shear mixer bottom driven	X	A, E
2	Mixer #2	High-shear mixer side driven	X	B, C
3	Dryer #1	Fluid bed dryer	X	A, B, C
5	Mill #1	Screening mill with oscillating bars	X	A, C
6	Mill #2	Screening mill with oscillating bars	X	B, E
8	Blender #1	Twin shell blender	X	A, D
9	Blender #2	Twin shell blender	2X	B, E
12	Tablet press #2	Force fed	X	B, E
13	Tablet press #3	Gravity fed	X	C, D

Table 7.6. Equipment and Products Grouping

Group No.	Equipment No.	Equipment Name	Operating Principle	Product Contact Surface Area	Products
1	1	Mixer #1	High-shear mixer bottom driven	Y	A, E
2	2	Mixer #2	High-shear mixer side driven	Y	B, C
3	3	Dryer #1	Fluid bed dryer	Y	A, B, C
4	5 6	Mill #1 Mill #2	Screening mills with oscillating bars	Y Y	A, B, C, E
5	8	Blender #1	Twin shell blender	25,000 cm^2	A, D
6	9	Blender #2	Twin shell blender	50,000 cm^2	B, E
7	12	Tablet press #2	Force fed	Y	B, E
8	13	Tablet press #3	Gravity fed	Y	C, D

Table 7.7. Equipment for Cleaning Validation

Group No.	Equipment No.	Equipment Name	Operating Principle	Product Contact Surface Area	Products
1	1	Mixer #1	High-shear mixer bottom driven	Y	A, E
2	2	Mixer #2	High-shear mixer side driven	Y	B, C
3	3	Dryer #1	Fluid bed dryer	Y	A, B, C
4	5	Mill #1	Screening mill with oscillating bars	Y	A, B, C, E
5	8	Blender #1	Twin shell blender	25,000 cm^2	A, D
6	9	Blender #2	Twin shell blender	50,000 cm^2	B, E
7	12	Tablet press #1	Force fed	Y	B, E
8	13	Tablet press #3	Gravity fed	Y	C, D

Finally, and as an additional equipment worst case, the hard-to-clean locations are identified for the representative piece of equipment in a group. The table in Chapter 5 may be used as a guide to identify the hard-to-clean locations for the most commonly used equipment for solid dosage forms. The hard-to-clean locations are then marked on the equipment diagram, which is included in the validation protocol.

The whole data collection and selection processes are summarized in Table 7.8, which represents the equipment database.

PRODUCT GROUPING

The aim of the product grouping is to determine the representative product manufactured in a piece of equipment. The representative product is the one containing the most insoluble active ingredient in the solvent used to clean the equipment. The contamination limit is computed by using the worst-case characteristics of all the products manufactured in the representative piece of equipment in an equipment group.

The characteristic data of the products manufactured in each equipment group determined previously are tabulated. These data, as explained in Chapter 6, are product related: active ingredient, solubility, *LTD*, daily dose and unit weight; and process related: batch size. If the product contains more than one active ingredient, all of them shall be listed. The data for the products manufactured in the equipment groups are presented in Table 7.9.

SELECTION OF THE PRODUCT WORST CASE

The worst case related to products is the most insoluble active ingredient. As defined in the cleaning validation policy, cleaning validation is performed after cleaning the equipment from the product containing the most insoluble active ingredient. The analytical detection method is developed for this ingredient. The solubility of the various active ingredients is listed according to the pharmacopeial definitions, i.e., very soluble, freely soluble, soluble, sparingly soluble, slightly soluble, very slightly, or insoluble. To differentiate between active ingredients of the same solubility definition, their solubilities must be given in mg/ml of water (The Merck Index, current edition). It can be seen in Table 7.9 that only products A, B, C, and E participate in the cleaning validation studies, since one of their active ingredients is the most insoluble. This also means that analytical methods are developed only for active ingredients "a", "b", "c", and "e".

As to the other characteristics of the products, the worst cases are selected, with the contamination limit equation in mind, to create the virtual all-worst-case product. The worst cases are represented by products with the lowest *LTD*, the highest maximal daily dose in units (D), the highest unit dose weight (Wt) and the smallest batch size (Wb).

The selected worst cases related to products in Table 7.9 appear in italic type. As an example, the virtual worst-case product in equipment group no. 5 has the characteristics listed in Table 7.10.

Table 7.8. Equipment Database

Group No.	Equipment No.	Equipment Name	Operating Principle Surface Area	Product Contact	Hard-to-Clean Locations	Products
1	1	Mixer #1	High-shear mixer bottom driven	Y	Lid gasket Discharge valve Impeller blades Chopper blades Vent filter	A, E
2	2	Mixer #2	High-shear mixer side driven	Y	*	B, C
3	3	Dryer #1	Fluid bed dryer	Y	*	A, B, C
4	5	Mill #1	Screening mill with oscillating bars	Y	*	A, B, C, E
5	8	Blender #1	Twin shell blender	25 000 cm^2	*	A, D
6	9	Blender #2	Twin shell blender	50 000 cm^2	*	B, E
7	12	Tablet press #1	Force fed	Y	*	B, E
8	13	Tablet press #3	Gravity fed	Y	*	C, D

* Refer to the table in Chapter 5

Table 7.9.   Product Database

Group No.	Product Name	Active Ingredient(s)	Solubility In Water (mg/ml)	LTD (mg)	Daily Dose (Units/day)	Unit Weight (mg)	Batch Size (kg)
1	A	*a*	*Insoluble*	*0.5*	4	170	238
	E	e	Very slightly	10	*8*	*245*	245
		g	Very soluble	100	8	245	245
2	B	*b*	*Slightly soluble*	*2*	*6*	123	247
	C	c	Sparingly soluble	125	3	*1060*	265
		f	Freely soluble	50	3	1060	265
3	A	*a*	*Insoluble*	*0.5*	4	170	238
	B	b	Slightly soluble	2	*6*	123	247
	C	c	Sparingly soluble	125	3	*1060*	265
		f	Freely soluble	50	3	1060	265
4	A	*a*	*Insoluble*	*0.5*	4	170	238
	B	b	Slightly soluble	2	6	123	247
	C	c	Sparingly soluble	125	3	*1060*	265
		f	Freely soluble	50	3	1060	265
	E	e	Very slightly	10	*8*	245	245
		g	Very soluble	100	8	245	245

continued on following page

continued from pevious page

Table 7.9.   Product Database

5	A	*a*	*Insoluble*	*0.5*	4	170	*238*
	D	d	Soluble	1	*8*	*530*	493
		h	Soluble	25	8	530	493
6	B	b	Slightly soluble	*2*	6	123	247
	E	*e*	*Very slightly*	10	*8*	*245*	*245*
		g	Very soluble	100	8	245	245
7	B	b	Slightly soluble	*2*	6	123	247
	E	*e*	*Very slightly*	10	*8*	*245*	*245*
		g	Very soluble	100	8	245	245
8	C	*c*	*Sparingly soluble*	*1*	3	*1060*	*265*
		f	Freely soluble	50	3	1060	265
	D	d	Soluble	125	*8*	530	493
		h	Soluble	25	8	530	493

In other words, the cleaning validation study of group no. 5 is performed on Blender #1 (equipment no. 8 in group no. 5) using the virtual worst-case product as the representative of product A (active ingredient "a") and product D (active ingredients "d" and "h"). The analytical detection method is developed for active ingredient "a", since it is the most insoluble of the active ingredients in the products manufactured in equipment group no. 5.

Following the same principles, a virtual worst-case product is obtained for each of the equipment groups.

CLEANING AGENTS

A similar approach is applied to the cleaning agents, using the LD50 of the most toxic component instead of *LTD* and the most insoluble component to represent the virtual worst-case cleaning agent.

For group no. 5, the cleaning agent characteristics are tabulated in Table 7.11.

CALCULATION OF THE CONTAMINATION ACCEPTANCE LIMIT

Active Ingredient

By putting the characteristics of the virtual worst-case product presented in the contamination acceptance limit equation and assuming a recovery factor of 70%:

$$MC \text{ (mg/swab)} = \frac{LTD/1000}{D} \times \frac{Wb}{Wt} \times \frac{Ss}{Se} \times R$$

we obtain the contamination acceptance limit of active ingredient "a" after cleaning equipment no. 8:

$$MC \text{ (mg/swab)} = \frac{0.5 \text{ mg}/1000}{8} \times \frac{238 \times 10^3 \text{ g}}{0.530 \text{ g}} \times \frac{25 \text{ cm}^2}{50\,000 \text{ cm}^2} \times 0.7$$

$$MC \text{ (active)} = 9.8 \times 10^{-3} \text{ mg/swab}$$

Cleaning Agent

The contamination acceptance limit for the cleaning agent is calculated by using the same equation and replacing LTD by LD50.

$$MC \text{ (mg/swab)} = \frac{LD50/1000}{D} \times \frac{Wb}{Wt} \times \frac{Ss}{Se} \times R$$

CLEANING VALIDATION MASTER PLAN

It has been seen that only products A, B, C, and E participate in the cleaning validation studies. Analytical methods must now be developed to detect active ingredients "a", "b", "c", and "e", at a range bracketing the respective calculated contamination limit for the

Table 7.10. Sample Worst-Case Product

Group No.	Equipment No.	Equipment Name	Product Name	Active Ingredient	Solubility In water (mg/ml)	LTD (mg)	Daily Dose D (units/day)	Unit Weight Wt (g)	Batch Size Wb (g)
5	8	Blender #1	Virtual product	a	Insoluble	0.5	8	0.530	238×10^3

Table 7.11. Cleaning Agent Characteristics

Group No.	Equipment No.	Equipment Name	Cleaning Agent	Active Ingredient	Solubility In water (mg/ml)	LTD (mg)	Virtual Product Daily Dose D (units/day)	Virtual Product Unit Weight Wt (g)	Virtual Product Batch Size Wb (g)
5	8	Blender #1	Kleen	k	Soluble	X	8	0.530	238×10^3

pieces of equipment participating in the cleaning validation studies. The range of calculated contamination limits of an active ingredient in the various equipment groups may be significantly different. In such cases, the analytical detection method is developed for the different ranges at the same time, thus saving important analytical resources.

At this stage, a cleaning validation master plan is drawn up and serves as the basis for the determination of all the activities to be performed to conduct the cleaning validation studies in a cost-effective and timely manner. The cleaning validation master plan is a comprehensive and controlled document that include the following sections:

- Cleaning validation policy
- Product and equipment database
- Product and equipment grouping
- Product and equipment worst cases
- Calculation of the contamination limit for the active ingredient and the cleaning agent for each group of equipment

All the data are consolidated in one table listing the critical equipment and product parameters to be taken into account to perform the cleaning validation studies (see Table 7.12).

WORKING PLAN

A detailed working plan, based on the cleaning validation master plan and in accordance with a priority order, is then prepared to include all the activities necessary to perform the cleaning validation study for each piece of equipment, the time for completion of each activity, as well as responsibilities. Table 7.13 presents the sequence and schedule of these activities.

EXECUTION

The execution of the cleaning validation studies includes all the activities listed in Table 7.13. These activities are delineated in the cleaning validation protocol specific for a piece of equipment representing the worst case of an equipment group, for example equipment no. 8 (Blender #1) representing equipment group no. 5.

CLEANING VALIDATION PROTOCOL

A basic protocol is written in a generic format and adapted to be specific for the piece of equipment to be validated. The protocol should refer to the cleaning validation master plan, be concise, and give clear instructions how to conduct the cleaning validation study. Any other applicable document should only be referred to in the text, available

Table 7.12. Cleaning Validation Master Plan

Group No.	Equipment No.	Equipment Name	Product Contact Surface Area	Hard-to-Clean Locations	Products	Virtual Product Active	Contamination Limit (active) mg/swab	Contamination Limit (cleaning agent) mg/swab
1	1	Mixer #1	Y	Lid gasket Discharge valve Impeller blades Chopper blades Vent filter	A, E	A (a)	* *	* *
2	2	Mixer #2	Y	*	B, C	B (b)	* *	* *
3	3	Dryer #1	Y	*	A, B, C	A (a)	* *	* *
4	5	Mill #1	Y	*	A, B, C, E	A (a)	* *	* *
5	8	Blender #1	25,000 cm²	*	A, D	A (a)	9.8×10^{-3}	* *
6	9	Blender #2	50,000 cm²	*	B, E	E (e)	* *	* *
7	12	Tablet press #1	Y	*	B, E	E (e)	* *	* *
8	13	Tablet press #3	Y	*	C, D	C (c)	* *	* *

* Refer to the table in Chapter 5

* * Calculate as in the example for group no. 5

Table 7.13. Working Plan

Activity	Responsibility	First Year				Second Year			
		Q1	Q2	Q3	Q4	Q1	Q2	Q3	Q4
Analytical development	R&D Lab	Active "a"	Active "c"	Active "e"	Active "b"				
Analytical method validation	R&D Lab	Active "a"	Active "c"	Active "e"	Active "b"				
Analytical method implementation	QC Lab	Active "a"	Active "c"	Active "e"	Active "b"				
Recovery study	QC Lab	Active "a"	Active "c"	Active "e"	Active "b"				
Validation protocol	Validation group		Equipment groups 1, 3, 4, 5	Equipment group 8	Equipment groups 6, 7	Equipment group 2			
Manufacture and sampling	Manufacturing and QA			Product A	Product C	Product E	Product B		
Testing of samples	QC Lab			Product A	Product C	Product E	Product B		
Validation report	Validation group				Equipment groups 1, 3, 4, 5	Equipment group 8	Equipment groups 6, 7	Equipment group 2	

for use at the time the study is conducted, and attached to cleaning validation report. The equipment specific protocol should be reviewed and approved by the validation group, production, and quality assurance before the execution of the study. Any deviation from the protocol must be investigated and the results of the investigation reviewed and approved by the same responsible persons. The protocol may be amended, when scientifically justified.

A cleaning procedure, work instructions for sampling and recovery, and swab characterization for a high-shear mixer are found in Part Three. Also presented in Part Three is a template of a cleaning validation protocol for a high-shear mixer. The fill-in-the-blank, easy-to-follow format is designed to be used for your own purposes.

The cleaning validation protocol should contain the sections listed in Table 7.14.

SAMPLING AND TESTING

Sampling is performed, as a worst case, at the end of the time limitation allowed before cleaning. To prevent material from drying, and subsequent difficulties in cleaning, this

Table 7.14. Cleaning Validation Protocol

- Purpose
- Responsibility
- Frequency
- Procedure
    - Manufacturing
    - Sampling
- Acceptance criteria
    - Visual inspection
    - Contamination acceptance limit
- Documentation
    - Product, cleaning agent, and equipment database
    - Cleaning procedure
    - Drawings
        - Equipment surface area
        - Hard-to-clean locations
- Analytical method and method validation

time period should be as short as feasible. Swab sampling is performed in accordance with the working instructions and an approved diagram of the equipment, marked with locations to be sampled. A swab sampling record is filled in by the sampler to document the actual sampling performance. It is important that the sampler be well trained in the swabbing technique and cognitive to report atypical events during sampling. In order to prevent adventitious contamination, powder-free gloves must be worn by the sampler. The swabs must be protected immediately after sampling in closed containers.

The residual materials in the swabs are extracted into a measured volume of an appropriate extraction solvent, preferably by sonication or by using a magnetic stirrer—as soon as possible after sampling and before testing—according to a standard operating procedure. When testing is not performed in a reasonable time period, the swab samples are kept in the extraction solvent under refrigeration. In such a case, the stability of the active ingredient in these conditions has to be determined.

After extraction, the solvent is filtered and tested according to the appropriate analytical method. The results are recorded in mg (or μg) per ml, converted to mg or (μg) per swab, and compared to the calculated contamination limit.

CLEANING VALIDATION REPORT

The cleaning validation report closely follows and refers to the cleaning validation protocol and includes the records of sampling and testing performance, as well as the testing results. In order to reduce paperwork, it is recommended to use a generic protocol with specific attachments relating to the equipment being validated. These attachments relate to the equipment and product grouping, worst cases, cleaning agent characteristics, all-worst-case virtual product, equipment diagram marked with the sample locations, equipment surface area, analytical method used, and the calculation of the acceptable contamination limit. The cleaning validation report ends with a conclusion as to the compliance of the results with the acceptance criteria.

8

Cleaning Validation
Monitoring and Maintenance

MONITORING

With all its limitations, the visual inspection of the equipment after cleaning and before use remains an essential part of the cleaning procedure. However, a periodic testing of residues may be necessary to confirm the reproducibility of the cleaning procedure. Clearly, the more automated the cleaning process, the less testing has to be performed.

For fully automated, well-designed, and properly validated CIP systems, no cleaning monitoring is needed. The reproducibility of the cleaning process is ensured though the proper functioning of the system and verification of the cleaning cycle parameters for each run, and is maintained through the change control procedure.

The reproducibility of the manual cleaning process cannot be ensured with the same degree of confidence as with automated systems. The control of manual cleaning resides in well-designed and detailed cleaning instructions and in intensive operator training. The role of the human factor has already been emphasized in Chapter 4. It is imperative to continuously enforce and monitor the performance of manual cleaning procedures. Consequently, periodic monitoring is required to confirm the reproducibility and the robustness of the cleaning procedure—when performed by the same operator or by different operators, at different times. Cleaning monitoring is generally performed by repeating the initial cleaning validation study, using the same sampling and analytical methods, and the same acceptance contamination limit. Since the sampling and testing activities involve various disciplines, they have to be coordinated and planned in advance and this may have an unwanted impact on operator diligence.

MAINTENANCE OF VALIDATION

The design and development of a pharmaceutical product does not necessarily require cleaning validation following the manufacture of the new product. However, the new product and its characteristics, the equipment in which it is manufactured, and the cleaning procedure are added in the database in order to determine whether the new product introduces a new worst case in any one of the parameters listed. If this is not the case, the pieces of equipment involved in the manufacture of the new product are considered to remain in a state of validation. Otherwise, a reevaluation of the state of validation must be made, depending on the new product characteristics introduced in the database. As one can see from the calculation of the contamination limit, lower values for the lowest therapeutic dose (*LTD*) or for the batch size (*Wb*) and higher values for the maximum daily dose (*D*) or for the unit dose weight (*Wt*) will give a lower maximum allowable contamination (*MC*) in mg per swab. The actual results obtained from the previous validation are compared to the new value of *MC*. If these results are lower than the new *MC*, the piece of equipment is considered to remain in a state of validation. If they are higher, the cleaning validation study has to be repeated after cleaning the piece of equipment in which the new product has been manufactured. The cleaning validation is performed by testing active ingredient residues from the new product, after having developed an analytical method sensitive enough to detect the new product at a level corresponding to the newly calculated contamination limit. Another case where the cleaning validation study has to be repeated is when the solubility of the active ingredient of the new product is lower than the solubility of any other active ingredient already listed.

The effectiveness of the cleaning procedures in relation to the product, the equipment, the cleaning procedure, the cleaning validation methods, or the cleaning monitoring results, must also be reevaluated in the following events:

Product

- The introduction of a new product

- Deletion of a product

- Any change in existing product, such as a change in formulation or in strength, or a smaller batch size

Equipment

- The introduction of a new equipment

- A change to another equipment

- A change in the equipment, such as the design, the materials of construction and a change in the product contact surface area

Cleaning Procedure

- The addition of a step in the cleaning procedure

- The deletion of a step in the cleaning procedure

- A change in the cleaning parameters (time, temperature, pressure, volume of solvent)

- A change in the cleaning agent

Cleaning Validation Methods

- A change in the sampling method

- A change in the testing method

The aforementioned changes are all governed by the change control procedure, a critical procedure established to review, approve, and evaluate the impact of the changes on the cleaning validation status. The rigorous implementation of the change control procedure, as it applies to cleaning validation, ensures the continuous compliance of the cleaning validation program. Following evaluation by R&D, engineering, production, and QA, the changes may be categorized as major or minor, with the major ones requiring either full revalidation of the cleaning procedure for three runs or partial revalidation for one or more runs, and the minor ones no revalidation. The assessment of the risks involved in the changes and the sound judgment of the potential impact of these changes must be done carefully in order not to compromise the equipment validation status, and must be documented in a formal manner.

The importance of the change control procedure, which covers every aspect of the pharmaceutical operations, cannot be stressed enough. This procedure provides a full mapping of the actual procedures and processes in the plant, and dissipates any doubt about what is really occurring and how changes may impact on product quality. As an example, a change in a pressure transmitter may result in changes in pressure differentials, and consequently in air-flow patterns, creating ideal conditions for cross contamination. An effective change control system should be in place so as to provide a powerful tool for knowledgeable and responsible personnel to assist them in the evaluation, and subsequent testing, of the impact of the change on quality. The rigorous implementation of the change control procedure is the key to prevent unexpected or unnoticed deviations and represents a fundamental aspect of the quality assurance system of the company.

A standard operating procedure "Cleaning Validation—Change Control" is presented in Part Three of this book.

9

Special Cleaning Validation Issues

BIOCONTAMINATION

Although focusing on chemical contamination, cleaning validation would not be complete without a reference to equipment biocontamination or bioburden. "Cleaning/sanitization studies should address microbiological contamination for APIs and intermediate processes for which microbiological contamination is a concern" (FDA, March 1998).

Nonsterile dosage forms are manufactured with materials and in environments prone to the proliferation of microorganisms. The presence of water and growth-supporting inactive ingredients—such as lactose and starch—in product formulations, the microbiological quality of the water as an ingredient or as the cleaning solvent, and the fact that most operations involve human intervention, impose the implementation of a control and monitoring program to limit the bioburden. Equipment design and the materials of construction have an important function in the minimization of bioburden. Dead legs or portions of the equipment, which may contain stagnant water, must be identified and eliminated wherever possible. Draining/drying of the equipment after cleaning and time limitation for the storage of equipment after cleaning are critical measures to minimize equipment bioburden.

Cleaning procedures must be evaluated and validated with the aim of demonstrating that the procedures effectively remove not only the active ingredient contaminant, but also viable microorganisms from the equipment product contact surfaces (Cooper 1998).

The microbiological methods used to determine bioburden are the same as used for monitoring sterile manufacturing equipment and environments. As opposed to the analytical detection methods for chemicals, these methods are straightforward and do not require a special development. However, a recovery study should be performed and the methods should be validated to show that no substance, such as residual active ingredient or cleaning agent, inhibits microorganism growth. If present, such residues must be neutralized prior to testing. Representative colonies of the microorganisms isolated during cleaning validation should be identified in order to build a plant microbial flora baseline with the aim of locating and eliminating potential contamination sources. Special attention should be given to the identification of objectionable microorganisms.

Two sampling methods are used for sampling equipment product contact surfaces.

Surface Sampling

Contact plates are used for sampling large smooth surfaces. This type of plate, commercially available, contains a sterile growth medium with an outward swelling surface. The growth medium, lightly pressed on the surface to be sampled, collects any microorganism present on that surface. The plate is incubated at the appropriate temperatures and time and colony forming units (CFU) representing the viable organisms are counted. Care must be taken to thoroughly clean the area sampled to remove residual growth medium from the equipment surface after sampling. Due to the intrinsic variability of microbiological determination, contamination levels rather than limits are set to indicate, if exceeded, that action is required.

Swabs, such as cotton or cotton tips, are also used for sampling surfaces, but are especially convenient to reach poorly accessible locations. Sterile swabs wetted with a sterile solution are rubbed over the surface to be sampled to remove microorganisms from the equipment surface and then rubbed again onto the surface of a plate containing a sterile solid growth medium to release the microorganisms collected. The plate is incubated as described above for contact plates.

The acceptance level for surface sampling is 10 CFU per 25 sq. cm.

Rinse Sampling

This sampling method is particularly convenient for equipment that cannot be disassembled easily, for large pieces of equipment, such as blenders, and to inaccessible locations or parts of the equipment, such as filling needles. After cleaning, a sample of the rinse solution is tested, by pouring onto a sterile growth medium plate or by filtration through a sterile membrane filter which is then laid on a sterile growth medium plate. The plate is incubated and CFUs are counted as above.

The acceptance level for rinse sampling—not more than 100 CFU per 1 ml—is the same as required by official compendia and acceptable to the FDA for Purified Water (FDA Guide to Inspections of Water Systems 1993).

PENICILLIN CONTAMINATION

In the case of penicillins, due to their allergenic potential, FDA's approach goes beyond the inclusion of a safety factor and requires no detectable level of penicillin in a drug suspected to be contaminated with penicillins, when tested according to a sensitive microbiological method. In the manufacture of penicillins, there is no reasonable way of decontamination and it is extremely difficult to attain a nondetectable level of residuals. Therefore, the use of dedicated facilities and equipment is recommended for this purpose.

There are some reports of the decontamination of facilities and equipment in order to convert them for use with nonpenicillin products. Although not recommended, such undertaking may be feasible when a systematic approach is employed. The difficulty resides in the fact that penicillins are easily disseminated due to their small particle size. On the other hand, penicillins are labile materials, easily decomposed in the presence of water. In order to increase the chance of success in a decontamination project, a 0.1 N sodium hydroxide solution in water is used. Addition of a surfactant, such as sodium lauryl sulfate, is useful to enhance the penetration of the sodium hydroxide solution in crevices and not readily accessible areas. The sodium hydroxide solution will not cause damage to stainless steel, but care should be taken to avoid contact with other metallic surfaces, such as aluminum. The recommended procedure is first to wash the walls, floors, and ceilings many times with plenty of water, then spray the decontamination solution. A time of contact should be allowed before a final rinse with water. The inactivation of penicillins in the way described has been reported by Allan and Deeks (1996) and Hakimipour (1984).

FDA's GMP Regulations 21 CFR Section 211.176 specifies the analytical method to be used for the detection of penicillins: "Procedures for Detecting and Measuring Penicillin Contamination in Drugs". This is a microbiological analytical method using a microorganism sensitive to penicillin. Another microbiological method widely used in the dairy industry to check the presence of penicillin in raw cow's milk makes use of *bacillus stearothermophilus*. A rapid method using this organism in a disk assay has been developed and validated to provide results after three hours of incubation at 55°C.

Penicillin decontamination seems therefore possible, with some caveats. First, the regulatory authorities must be consulted and must agree with the overall project plan and outcome. Second, a monitoring program for testing residual penicillin in the environment, as well as in the products, is mandatory over a period of time.

NONPHARMACEUTICALS

The use of common equipment for the manufacture of APIs with nonpharmaceuticals, such as pesticides and herbicides, has been addressed in FDA's Human CGMP Note, June 1998.

> "Some nonpharmaceuticals pose unacceptable risks of cross contamination. . . . In some cases, in addition to separate equipment, it would be appropriate to use a separate facility. . . . These risks are influenced by the nature and intended use of the drug product that will incorporate the API. . . . Those risks may be of greater concern [for] large doses, long term therapy, treating open wounds, injection or inhalation."

CONTRACT MANUFACTURING AND PACKAGING

Contract manufacturers and packagers are subject to the same regulations and must meet the same requirements related to cleaning validation as the companies that require their services. However, it should be kept in mind that the regulatory authorities regard the company that holds the marketing authorization as responsible for the safety, identity, and purity of the marketed product. It is therefore the company's duty to check and verify that the contract manufacturer or packager has a cleaning validation program in place. The matter is complicated by the fact that these contractors manufacture or package for many other companies and the contract giver cannot get the knowledge of the safety and cleanability of the other products, due to secrecy agreements. On the other hand, the contracting business is based on giving the service and quality in the conditions in which they operate. This is their *raison d'etre* and they would be out of business if they did not implement a cleaning validation program. Nevertheless, the difficulty when dealing with contractors resides in the fact that their validation program would probably be different from yours. Understand their cleaning validation rationale and verify the implementation of the cleaning validation program. Audit raw data of the cleaning validation testing before your product is manufactured, where all confidential and trade secrets have been blacked out. Check that a procedure is in place when a new product is introduced to make the necessary corrections to their contamination acceptance limit. Another approach is to require your contractor to comply with your company's cleaning validation policy. A third approach is to require the contractor to perform a cleaning verification of the equipment before use with your product, using an absolute limit as described in Chapter 5.

10

Cleaning Validation Trends

INTRODUCTION

"What's the future of cleaning validation?" While it is difficult to be a prophet in the regulatory field, we may make the unworthy prediction, based on the experience gained with other regulatory issues, that cleaning validation is here to stay. The regulatory requirements regarding cleaning validation would have forced the industry to look again at cleaning procedures as an important way to ensure patient safety.

The resources allocated to cleaning validation represent an additional financial burden and a significant investment of time, which otherwise could be allocated to the research and development of new products. The approach proposed in this book will alleviate the burden imposed on companies by making cleaning validation manageable and cost effective. Yet, due to constraints related to the sampling and testing methods, one has to be creative and explore new ways to achieve, in an efficient way and in a shorter time, the purpose of ensuring patient safety. Following is a review of developments, some already available and one futuristic, which would provide the same safety assurance level and additional operational benefits.

CONTAMINATION CONTROL

As we have shown in Chapter 1, contamination control is achieved by taking a series of measures, after the manufacturing processes have been analyzed, to locate the sources of contamination and implement procedures to prevent or at least minimize them. The manufacturing process, the personnel, the equipment, and the cleaning procedure all may contribute to the contamination carryover.

PERSONNEL

The role of the operator in ensuring the cleanliness of equipment has been emphasized in Chapter 4. If it is crucial when using manual cleaning methods, it is not less critical with automated cleaning.

EQUIPMENT

Since the pharmaceutical industry is becoming more competitive while regulatory requirements are becoming tougher, equipment manufacturers are expected to design and construct machines which can fulfill the demand for economical cleaning performance and yet give results of the desired quality (Faubel et al. 1990; Parikh 1998; Hausmann et al. 1998; Baseman 1992).

Automation in manufacturing coupled with automation in cleaning, such as CIP, can bring significant savings. In a conventional tablet press, the degree of dismantling is high to allow cleaning of the hard-to-clean locations. Consequently, the cleaning process is lengthy, a hardly acceptable fact in a multiproduct facility that can discourage frequent changeover between different products, thus turning flexibility into wishful thinking only.

A tablet machine equipped with a CIP system and designed with these aspects in mind has been developed (Hausmann et al. 1998). In this machine, the punches are removed and then the CIP starts. Fluid bed dryers have been developed to ensure conformance to the user's predetermined cleaning validation criteria. These pieces of equipment include new features such as CIP stainless steel filters instead of the classical fabric filters which have to be dismantled and cleaned separately, sanitary and flush fittings to prevent agglomeration of contaminants, and inflatable gaskets made of inert material to prevent dust propagation. The cooperation between equipment manufacturers and pharmaceutical companies will bring further developments.

It is expected that because of demand for CIP systems for complicated machines such as tableting, encapsulating, and filling machines, the equipment design as well as the design of the manufacturing areas will change. Only product contact parts of the equipment will be located in the clean areas, while the machinery parts will be located in sufficiently spacious technical areas, adjacent to the clean areas. With CIP systems and washing machine for small equipment parts and ancillary equipment, space allocated for washing could be reduced significantly.

Closed systems, whereby materials cannot contaminate the environment and are not exposed to it during their transfer from a piece of equipment to another, already exist. The application of the barrier isolation technology for oral solid dosage forms is in its infancy and is expensive, but has a promising future.

CLEANING PROCEDURE

As mentioned before, CIP systems and washing machines will replace human labor. However, the control and maintenance of the machine will still be performed by human beings. Machines are not exempt from failures, and procedures shall be in place, including alarms, to prevent and report malfunctioning.

CLEANING AGENTS AND TOOLS

The cleaning agent plays a critical role in manual or automated cleaning methods. The importance of the cleaning agent is threefold: to perform efficient cleaning, to be nontoxic to the patients, and to be environment friendly. In addition, the cleaning agent must be detectable by usual analytical methods and must not interfere in the analytical procedure of the active ingredient. These attributes form the basis for the development of new cleaning agents. The cooperation between cleaning agent manufacturers, pharmaceutical companies, and the environmental agencies is essential to develop the ideal cleaning agent.

SAMPLING TECHNIQUE

The swab sampling technique, preferred by the regulatory agencies, will remain an issue. This technique is demanding, operator dependent, and requires well-trained operators. As explained in Chapter 5, the rinse water technique, which is more rapid and much more simple to use, is not acceptable. It is not expected that regulatory authorities will change their mind on this issue. The only way to apply the swabbing technique in a consistent manner is to give very detailed swabbing instructions and to train the operators.

The swabbing technique may be correlated to the detection method. Analytical method developments may radically change the picture.

ANALYTICAL METHODS

The analytical detection method must be specific and sensitive. These are rational requirements, but the presently acceptable methods are time consuming and instrumentation is expensive. The use of shorter HPLC columns alleviate these concerns in some cases. The analytical detection methods should ideally be rapid, specific, and sensitive. To achieve this purpose, new methods have to be developed. It may seem futuristic to envision the use of cleanliness sensors, similar in a way to the development of the near infrared sensors for the identification and quantitation of active ingredients. Yet, to the authors's minds, it seems to be the only kind of analytical technique, which would obviate the need for swabbing. The development of such sensors will create a breakthrough in cleaning validation and the authors urge analytical chemists and analytical instrument manufacturers to explore this possibility, for the benefit of all.

REGULATORY ISSUES

As shown in the discussion of the contamination limit, there is no consensus on which considerations the limit should be calculated. Although the rationale to base the calculation on toxicological effect seems to be extensively applied, there is clearly a need for the regulatory agencies or for nonprofit pharmaceutical organizations to provide more specific requirements. Now that experience has accumulated in the industry as well as in the regulatory agencies, there is more insight and understanding on what the acceptance criteria should be. As for other hot regulatory issues, it takes time to assimilate and digest the experience gained in the industry. It happened more than once that the industry itself shaped regulatory requirements, by publicly offering their know-how in conferences and industry agency meetings. It is hoped that more specific guidance will be issued by regulatory agencies in the near future to clarify their expectations. Moreover, in our free trade world, harmonization of regulatory requirements is the word of the day and an immediate must in the field of cleaning validation.

THE FUTURE

There is a saying, "The future is not a sure thing", and that is why there are mixed feelings of apprehension and expectation about it.

As the future relates to cleaning validation, the bad news is that cleaning validation is here to stay. Pharmaceutical companies that did not start a cleaning validation program must address the issue without delay. For companies which have already implemented a cleaning validation program, the huge initial load is over. But they are still dealing with this issue during regulatory agency inspections and repeating parts of the cleaning validation studies for monitoring purposes or when introducing new products in their manufacturing lines.

The good news is not for the very near future. Yet, it is probable that once the regulatory agencies have enforced the implementation of cleaning programs in the industry, they will issue specific guidance and may ease their requirements. The real good news resides in the development of advanced analytical detection techniques, as mentioned, which will make a revolution in the field of cleaning validation. When this dream is realized, cleaning validation will be reduced to a simple analytical exercise, just as TOC has simplified the testing of purified water and water for injection in the pharmacopeial monographs. Cleaning monitoring could then be performed as an integral part of the manufacturing process and give continuous assurance of the equipment cleanliness level. What a wonderful dream, let us make it true!

Appendices

Cleaning and Cleaning Validation Procedures

Appendix A

Cleaning

POLICY

PROCEDURE

Policy

EQUIPMENT AND PROCESSING AREA CLEANING

The Rx Pharmaceutical Company	POLICY	
Edition No.:		Validation Unit:
Supersedes:	**Equipment and Processing Area Cleaning**	Production:
Effective Date:		Quality Assurance:

page 1 of 4

Purpose

Manufacturing equipment and processing areas should be cleaned to avoid contamination and cross contamination.

Responsibility

Quality assurance is responsible for drawing up a policy concerning cleaning processes in accordance with GMP regulations.

R&D and production are responsible for developing cleaning procedures and standard operating procedures in accordance with this policy.

Operators, supervisors, and managers are responsible for working according to cleaning procedures and standard operating procedures relating to this policy.

Quality assurance is responsible for verifying the proper implementation of the policy, cleaning procedures, and related standard operating procedures.

Frequency

Routinely, in accordance with the level of cleaning, defined as follows.

The Rx Pharmaceutical Company	POLICY	
Edition No.:		Validation Unit:
Supersedes:	**Equipment and Processing Area Cleaning**	Production:
Effective Date:		Quality Assurance:

Procedure

Level of Cleaning

Two levels of cleaning are defined for manufacturing equipment and processing areas: minor and major.

    a. Minor Cleaning

        Minor cleaning is defined as cleaning which is conducted between the manufacture of

- Batches of the same product

- Batches of different strengths of the same product

    b. Major Cleaning

        Major cleaning is defined as cleaning which is conducted between the manufacture of

- Different products

- A colored product followed by a white product with the same active ingredient

- For equipment cleaning after plant shutdown and when resuming work

Time Limitations

    a. Time between end of manufacturing and start of cleaning is set as follows:

- For solid oral dosage forms: 3 calendar days

- For liquid and semisolid oral and topical nonsterile dosage forms: 1 calendar day
  Interruptions for periods longer than these time limits require a major cleaning. For products which require shorter time limits, their specific cleaning procedure should indicate the appropriate time limits.

    b. Time between final rinse and drying should be kept to a minimum. Preferably, the equipment is dried immediately after final rinse.

The Rx Pharmaceutical Company **POLICY**		
Edition No.:	**Equipment and Processing Area Cleaning**	Validation Unit:
Supersedes:		Production:
Effective Date:		Quality Assurance:

c. Time until major cleaning is conducted while manufacturing product in a campaign

- For solid oral dosage forms: 14 calendar days

- For liquid and semisolid, oral and topical nonsterile dosage forms: 7 calendar days

d. Time until major cleaning is conducted for unused cleaned equipment: 30 calendar days.

Cleaning Procedure

Major cleaning of manufacturing equipment and processing areas are carried out using a specific "Equipment/Processing Area Cleaning Procedure". Minor cleaning is carried out according to a general departmental procedure.

Detailed Procedures for a Major Cleaning

- Manual cleaning

  a. Safety precautions to the operator

  b. The approved cleaning solvents, materials, and tools

  c. Degree of disassembly of the equipment and the hard-to-clean locations

  d. Washing steps

  e. Rinsing steps—final rinse is always conducted using purified water

  f. Drying

  g. Visual inspection and quantitative monitoring (if needed)

  h. Reassembly

  i. Storage conditions: closed, covered, storage room classification

- Automated cleaning

  An automated cleaning procedure should contain a cleaning sequence. Usually the relevant program contains all the relevant steps, such as: prewash, wash, rinse, final rinse, and drying, which run automatically. The sequence should detail the

The Rx Pharmaceutical Company	POLICY	
Edition No.:	**Equipment and Processing Area Cleaning**	Validation Unit:
Supersedes:		Production:
Effective Date:		Quality Assurance:

page 4 of 4

controlled process parameters, such as: temperature in each step, time, concentration of the cleaning agent, volume of water, flow rate, and mixing speed.

Visual Inspection and Monitoring

a. At the end of a major cleaning and prior to assembly, the equipment cleanliness is visually inspected and approved by the supervisor. Prior to manufacturing, cleanliness is verified by visual inspection and recorded by the operator in the manufacturing instructions of the next product to be manufactured.

b. For equipment which is periodically or routinely monitored for quantitative residues, a control and follow-up system to manage the monitoring must be in place. This monitoring will be carried out after cleaning and visual inspection.

Documentation

Specific Cleaning Procedure for Each Equipment/Processing Area

Equipment Storage

a. For equipment which is permanently located with all its ancillaries in a specific processing area, a single "Equipment/Processing Area Use and Cleaning Log" is used (Attachment I).

b. For equipment which is located in a storage room and used in different rooms, separate documentation has to be kept, both for the processing area (Attachment II) and the equipment (Attachment III).

Equipment Labelling

A "cleaned" label (Attachment IV) is attached to the equipment and/or the room or both.

Company Name: _____ Form _____ Ed. _____
Cleaning Procedure No.: _____

Equipment/Processing Area Use and Cleaning Log

Processing Area Name: _____ ID No: _____
Equipment Name(s):   _____ ID No: _____
                      _____ ID No: _____
                      _____ ID No: _____
                      _____ ID No: _____
                      _____ ID No: _____

Date	Product Name	Batch. No.	Cleaning Level Minor/Major	Name and Signature	
				Performed by	Approved by

Company Name:_____ Form _____ Ed. _____

Cleaning Procedure No.: _____

Processing Area Use and Cleaning Log

Processing Area Name: _____ ID No: _____

Date	Product Name	Batch. No.	Cleaning Level Minor/Major	Name and Signature	
				Performed by	Approved by

ATTACHMENT III

Company Name: _____ Form _____ Ed. _____

Cleaning Procedure No.: _____

Equipment Use and Cleaning Log

Equipment Name: _____ ID No: _____

Date	Product Name	Batch. No.	Cleaning Level Minor/Major	Name and Signature	
				Performed by	**Approved by**

ATTACHMENT IV

Company Name: _____ Form _____ Ed. _____

CLEANED

Product Name: _____
Batch No.:   _____
Start Date: _____

Equipment Name: _____
Cleaning Date: _____

                  Signature: Date:
Performed by: _____   _____
Checked by: _____   _____

Previous Product: _____
End of Work Date: _____

Cleaning Procedure

HIGH-SHEAR MIXER (DIOSNA)

The Rx Pharmaceutical Company	*CLEANING PROCEDURE*	
Edition No.:	**High-Shear Mixer (Diosna)**	Validation Unit:
Supersedes:	**Diosna I**	Production:
	Diosna II	
Effective Date:	**Diosna III**	Quality Assurance:

Equipment/Processing Area No.:	Previous Product:	Batch No.:	Date End:
_____ _____	_____	_____	_____

page 1 of 5

Safety Precautions: Wear gloves and goggles.
Materials: Detergent No. 1 for equipment
    Detergent No. 2 for floors and walls
    Detergent No. 3 for glass separations
Tools: Sponge
    Soft hair brush
Hard-to-clean locations: See attached drawing of the equipment.

Equipment

1. Prewash

    1.1 Remove all identification labels of previous product.

    1.2 Connect the equipment to the drain system.

    1.3 Wash the body and lid inside and outside with tap water while the equipment outlet is open to the drain.

2. Wash

    2.1 Close the equipment outlet to the drain.

    2.2 Fill with hot tap water up to 20% of its full capacity.

    2.3 Add 3 liters of Detergent No. 1.

    2.4 Close the lid.

    2.5 Operate for 15 minutes, Mixer Speed II; Chopper Speed II.

    2.6 Drain.

The Rx Pharmaceutical Company	CLEANING PROCEDURE	
Edition No.:	**High-Shear Mixer (Diosna)** **Diosna I** **Diosna II** **Diosna III**	Validation Unit:
Supersedes:		Production:
Effective Date:		Quality Assurance:

3. Disassembly

   3.1 Remove the sleeve.

   3.2 Remove the impeller and put it in a parallel position.

   3.3 Wash the sleeves, scrub the impeller and underneath the impeller and the chopper with the detergent solution and the brush.

4. Rinse
   Rinse with hot tap water.

   • Sleeve

   • Impeller

   • Body and lid, inside and outside

   • Outlet pipe

5. Final Rinse
   Rinse with purified water.

   • Sleeve

   • Impeller

   • Body and lid, inside

   • Outlet pipe

6. Dry

   6.1 Attach the hot air hose to the equipment chimney.

   6.2 Dry for 20 minutes while the equipment outlet is open.

Ancillary Equipment

1. Wash

   1.1 Dilute 2.5 liters of detergent No. 1 in 7.5 liters of tap water.

The Rx Pharmaceutical Company	**CLEANING PROCEDURE**	
Edition No.:	**High-Shear Mixer (Diosna)** **Diosna I** **Diosna II** **Diosna III**	Validation Unit:
Supersedes:		Production:
Effective Date:		Quality Assurance:

1.2 Use sponge and diluted detergent to clean.

- Hoses (used for water, granulation solutions or pastes, vacuum hoses)

- Ladder and stand

- Mixer and stainless steel container used for granulation solution or paste preparations

1.3 Fill with hot tap water 1/2 capacity of the stainless steel container and add detergent No. 1 to 1/8 capacity of the container. Operate the mixer for 10 minutes on full speed to clean the mixer.

2. Rinse
Rinse with tap water.

3. Final Rinse
Final rinse with purified water.

Process Area

1. Wash

1.1 Dilute 2.5 liters detergent No. 2 in 7.5 liters of tap water.

1.2 Use a brush to clean the ceiling, walls, floor, and the glass separators.

2. Rinse
Rinse the ceiling, walls, floor, and the glass separators with tap water.

3. Dry
Use a fabric to dry the surfaces.
Use detergent No. 3 and dry fabric to shine the glass separators.

The Rx Pharmaceutical Company	CLEANING PROCEDURE	
Edition No.:	**High-Shear Mixer (Diosna)** **Diosna I** **Diosna II** **Diosna III**	Validation Unit:
Supersedes:		Production:
Effective Date:		Quality Assurance:

Visual Inspection

If clean, mark "✓" in the box. If not clean, mark "x" in the box.

a. EQUIPMENT
- Body—inside ☐
- Lid—inside ☐
- Chopper—blades and underneath ☐
- Impeller—blades and shaft ☐
- Outlet pipe ☐
- Equipment—outside ☐

b. ANCILLARY EQUIPMENT
- Hoses ☐
- Ladder and stand ☐
- Mixer and container ☐

c. PROCESS AREA
- Ceiling ☐
- Walls ☐
- Glass separations ☐
- Floors ☐

Reassembly

If the equipment is clean, reassemble to allow proper function.   ☐

Status Label

Fill the "Clean" label with the relevant details. Attach the label to the clean equipment.   ☐

Storage Conditions

Cover the equipment.   ☐

The Rx Pharmaceutical Company	CLEANING PROCEDURE	
Edition No.:	**High-Shear Mixer (Diosna)**	Validation Unit:
Supersedes:	**Diosna I** **Diosna II**	Production:
Effective Date:	**Diosna III**	Quality Assurance:

page 5 of 5

Summary

The equipment, ancillaries, and the process area are all clean and dry.

Signature: _____ Date: _____

Manager's Signature: _____ Date: _____

Figure 11.1. Drawing of the Equipment

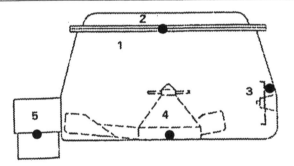

1. Body
2. Lid
3. Chopper
4. Impeller
5. Discharge chute

Hard-to-clean locations are marked with a bullet (·)

Documentation

Document the cleaning in the "Equipment/Processing Area Use and Cleaning Log" Form.

Appendix B

Cleaning Validation

POLICY

PROCEDURE

Policy

CLEANING VALIDATION

The Rx Pharmaceutical Company	POLICY	
Edition No.:		Validation Unit:
Supersedes:	**Cleaning Validation**	Production:
Effective Date:		Quality Assurance:

Purpose

The objective of cleaning validation is to attain documented evidence which provides a high degree of assurance that the cleaning procedure can effectively remove residues of a product and of the cleaning agent from the manufacturing equipment to a level that does not raise patient safety concerns.

Responsibility

1. Validation unit: Prepare validation protocol, conduct the study, and prepare a final report.

2. Production: Approve the validation master plan and participate in the study as required.

3. Engineering: Provide drawings with calculation of product contact area.

4. Quality assurance: Approve the validation master plan, the validation protocol, and the validation report.

5. Analytical R&D: develop and validate the analytical method.

6. Quality control: test samples.

The Rx Pharmaceutical Company	POLICY	
Edition No.:		Validation Unit:
Supersedes:	**Cleaning Validation**	Production:
Effective Date:		Quality Assurance:

Frequency

Cleaning Validation Study

On the worst-case piece of each group of equipment.
On three consecutive major cleanings of batches of the chosen worst case product.

Maintenance of Validation

The effect of changes on the effectiveness of the cleaning procedure should be evaluated and when required, a revalidation study on 3 consecutive batches after major cleaning (on different batches) will be performed. The changes can be related to

- Product

- Equipment

- Cleaning procedure

- Cleaning validation methods

Monitoring:

Once a year after a major cleaning of one batch on each validation group of equipment.

Procedure

The cleaning validation policy is based on

- Selecting the worst-case equipment

- Selecting the worst-case product

- Calculating the contamination acceptance limits for the active ingredient and for the cleaning agent

- Selecting the sampling method

- Selecting and developing the analytical method

The Rx Pharmaceutical Company	POLICY	
Edition No.:		Validation Unit:
Supersedes:	**Cleaning Validation**	Production:
Effective Date:		Quality Assurance:

page 3 of 5

Selecting the worst-case equipment

- Equipment grouping which is based on the following criteria:

    - Identical, interchangeable pieces of equipment with the same cleaning procedure.

    - Equipment with the same operating principles and the same cleaning procedure, but with different product contact surface area.

- The worst case is represented by the equipment with the larger surface area. The products matrix that should be considered is the matrix, which includes all the products manufactured in all the pieces of equipment in the group.

- The cleaning validation studies should take into account the hard-to-clean locations in each selected piece of equipment, in addition to the main parts of the equipment.

- Cleaning validation on dedicated equipment includes studies on residual cleaning agents only. This also includes dedicated fluid bed dryer filter bags.

Selecting the worst-case product

A database containing product characteristics and manufacturing parameters such as: product name, active ingredient, dose weight, batch size, solubility, LTD, and daily dose should be created for each equipment group. The worst-case product is selected from this database based on the lowest solubility of the active ingredient in water (if the cleaning procedure is water based), or in the relevant cleaning solvent.

Calculating the contamination acceptance limits for the active ingredient and for the cleaning agent:

- Contamination limit for the active ingredient
  The maximum allowable amount of contaminant (MC) in mg per swab, for a specific piece of equipment is calculated by selecting the worst-case values for parameters listed for the products in the matrix.

<table>
<tr><td colspan="2" align="center">***The Rx Pharmaceutical Company***</td><td align="center">***POLICY***</td></tr>
<tr><td>Edition No.:</td><td rowspan="3" align="center">**Cleaning Validation**</td><td>Validation Unit:</td></tr>
<tr><td>Supersedes:</td><td>Production:</td></tr>
<tr><td>Effective Date:</td><td>Quality Assurance:</td></tr>
</table>

$$MC \text{ (mg/swab)} = \frac{LTD/1000}{D} \times \frac{Wb}{Wt} \times \frac{Ss}{Se} \times R \qquad \text{Equation 1}$$

LTD	= Lowest therapeutic dose		(mg)
D	= Highest maximal daily dose		(dose units)
Wb	= Smallest batch size		(g)
Wt	= Highest unit dose weight		(g)
Ss	= Swab area		(cm^2)
Se	= Equipment product contact surface area		(cm^2)
R	= Recovery factor of active ingredient with lowest solubility		(%)

• Contamination limit for the cleaning agent:

$$MC \text{ (mg/swab)} = \frac{LD50/1000}{D} \times \frac{Wb}{Wt} \times \frac{Ss}{Se} \times R \qquad \text{Equation 2}$$

LD50 = Lethal dose of 50% of animal population   (mg/kg)

The sampling method

The cleaning validation studies are based on the swabbing technique. The selected swabs should be characterized for their suitability and a recovery test should be performed prior to the validation studies. The validation studies include swabbing of the main surface area as well as the hard-to-clean locations. Swabbing is carried out for the active ingredient and the cleaning agent.

The testing method

Validated selective and sensitive analytical methods should be used to determine the active ingredient and cleaning agent residues.

Acceptance criteria

a. The equipment should be inspected and found to be visually clean.

b. The maximum allowable amount of active ingredient contaminant in a piece of equipment cleaned with the cleaning procedure should be not more than the limit calculated in equation 1.

The Rx Pharmaceutical Company	**POLICY**	
Edition No.:		Validation Unit:
Supersedes:	**Cleaning Validation**	Production:
Effective Date:		Quality Assurance:

c. The maximum allowable amount of cleaning agent residue should be not more than the limit calculated in equation 2.

d. For maximum allowable amount of active ingredient and/or cleaning agent which can be visually inspected, a lower limit should be set so that at the new limit no contaminant can be observed.

Corrective action

In case of failure to demonstrate values as calculated for the contaminant limits, an investigation into the cause of failure should be conducted. Considerations such as improving the cleaning procedure and revalidation should be considered.

Standard Operating Procedure

CLEANING VALIDATION

The Rx Pharmaceutical Company	STANDARD OPERATING PROCEDURE	
Edition No.:		Validation Unit:
Supersedes:	**Cleaning Validation**	Production:
Effective Date:		Quality Assurance:

page 1 of 5

Purpose

The objective of cleaning validation is to attain documented evidence which provides a high degree of assurance that the cleaning procedure can effectively remove residues of a product and of the cleaning agent from the manufacturing equipment to a level that does not raise patient safety concerns.

Responsibility

Validation unit

- Prepare validation master plan, working plan, and protocol
- Calculate the contamination limit for the active ingredient and the cleaning agent
- Conduct validation study including sampling
- Prepare validation report
- Approve validation study

Production

- Approve validation master plan, working plan, and protocol
- Verify accuracy of the cleaning procedure
- Identify the equipment hard-to-clean locations
- Perform cleaning
- Approve validation study

Production planning

- Provide all information to build database

The Rx Pharmaceutical Company	STANDARD OPERATING PROCEDURE

Edition No.:		Validation Unit:
Supersedes:	**Cleaning Validation**	Production:
Effective Date:		Quality Assurance:

Engineering

- Verify accuracy of drawings and calculate product contact area

- Assist in the identification of equipment hard-to-clean locations

Quality assurance

- Approve validation master plan, working plan, and working protocol

- Approve the validation study

R&D analytical development

- Develop and validate the analytical method

Quality control

- Perform the recovery study

- Test samples and prepare analytical report

Frequency

- The cleaning validation is performed on each piece of equipment which was selected as representative of its group according to the equipment list

- The cleaning validation studies are conducted after three consecutive major cleanings of the selected product

- Maintenance of validation and monitoring are performed as described in the "Cleaning Validation—Policy"

Procedure

The equipment list should include all the equipment which is used in the manufacturing plant. The equipment should be listed according to the technology and grouped according to the principles listed in the "Cleaning Validation Policy".

The Rx Pharmaceutical Company	*STANDARD OPERATING PROCEDURE*	
Edition No.:		Validation Unit:
Supersedes:	**Cleaning Validation**	Production:
Effective Date:		Quality Assurance:

The validation unit prepares the specific equipment cleaning validation protocol for the relevant equipment group including:

a. Review and attach the equipment list

b. Review and attach of the cleaning procedure of the equipment

c. Attach the equipment diagram prepared by the engineering department. List the parts of the selected equipment.

d. Attach the drawings of the parts of the equipment and the calculated surface area of the parts as prepared by the engineering department, including:

   • Sampling locations, including the hard-to-clean locations

   • Swabbing area for the active ingredient

   • Swabbing area for the cleaning agent

   • Swab type used for sampling of the active ingredient

   • Swab type used for sampling of the cleaning agent

The validation unit reviews the product list and makes sure that it is updated and contains all the following parameters for all the products manufactured on the pieces of equipment included in the relevant equipment group.

a. The parameters in the list include:

   • Product name

   • Active ingredient in the product

   • Solubility of the active ingredient in water (mg/mL)

   • Lowest therapeutic dose (LTD) (mg)

   • Batch size (g)

   • Dose weight (g)

   • Daily dose (maximum units/day)

The Rx Pharmaceutical Company	STANDARD OPERATING PROCEDURE	
Edition No.:		Validation Unit:
Supersedes:	**Cleaning Validation**	Production:
Effective Date:		Quality Assurance:

b. The worst-case product due to the lowest solubility of its active ingredient in water is selected from the product matrix.

c. The cleaning agent used in the cleaning procedure is listed.

The validation unit attaches the following documents to the specific protocol:

a. The analytical method for quantitative determination of the active ingredient.

b. The analytical method validation report of the quantitative determination of the active ingredient.

c. The analytical method for quantitative determination of the cleaning agent.

d. The analytical method validation report for the quantitative determination of the cleaning agent.

e. The sampling recovery study report performed according to the working instructions. Record the sampling recovery yield in the specific protocol.

f. The swab characterization study according to the working instructions.

The validation unit calculates the maximum allowable amount of active ingredient and cleaning agent (*MC*) in mg per swab, for the piece of equipment according to the following equation:

$$MC(\text{mg/swab}) = \frac{LTD/1000}{D} \times \frac{Wb}{Wt} \times \frac{Ss}{Se} \times R$$

where the parameters are selected from the product matrix as follows:

a. The lowest *LTD* for the active ingredient or *LD50* for the cleaning agent

b. The highest maximal daily dose (*D*)

c. The smallest batch size (*Wb*)

d. The highest unit dose weight (*Wt*)

The Rx Pharmaceutical Company	STANDARD OPERATING PROCEDURE	
Edition No.:		Validation Unit:
Supersedes:	**Cleaning Validation**	Production:
Effective Date:		Quality Assurance:

The equipment cleaning validation specific protocol is prepared and signed by the validation unit. It is then approved by the production manager and the quality assurance manager.

After production of the virtual worst-case product in the worst-case piece of equipment selected from the group equipment, major cleaning of the equipment is executed by the production worker using the cleaning procedure. The cleaning is approved by the supervisor as stated in the cleaning procedure. When the equipment is approved as clean, the validation unit will inspect the equipment for visual cleanliness, including all components of the equipment as listed in the protocol—"Visual Inspection and Sampling". The visual inspection is documented on this form.

The listed parts are sampled both for the active ingredient and the cleaning agent—from different locations—as described in the working instructions. The sampling is documented in the "Visual Inspection and Sampling" form.

Any deviations or unusual events are documented under "Remarks".

All the samples are transferred immediately after sampling to the analytical laboratory for analysis.

This procedure is repeated after two additional and consecutive major cleanings of the same piece of equipment after manufacturing the worst-case product.

The analytical laboratory tests the samples using the relevant analytical method to determine the residual amounts of the active ingredient and of the cleaning agent per swab.

The validation coordinator summarizes all results in a cleaning validation report.

Documentation

Cleaning validation protocol and report.

Standard Operating Procedure

CLEANING VALIDATION—CHANGE CONTROL

The Rx Pharmaceutical Company	STANDARD OPERATING PROCEDURE	
Edition No.:		Validation Unit:
Supersedes:	**Cleaning Validation— Change Control**	Production:
Effective Date:		Quality Assurance:

page 1 of 1

Purpose

The purpose of this document is to define a procedure which ensures that all the changes which might affect equipment cleaning are reviewed and evaluated in relation to equipment cleaning validation.

Responsibility

Manufacturing and engineering departments are responsible for informing the validation unit of any changes related to equipment cleaning validation.

The validation unit is responsible for filling out the Change Control Form, change the database and recalculate the MC limits when required, and to evaluate the effect of the change on the status of the equipment cleaning validation for each relevant equipment.

Quality Assurance is responsible for approving the changes in the databases and the decisions related to revalidation.

Frequency

When a change is made.

Procedure

Changes in products, equipment, cleaning procedures, or cleaning validation methods may affect the status of the equipment cleaning validation. On any change related to these subjects, as listed in the "Decision Tree" (Attachment I), production and engineering departments should inform the validation unit.

The validation unit reports the change and categorizes it according to the list detailed in the "Decision Tree" on the "Cleaning Validation—Change Control Form" (Attachment II). The validation unit also details the change.

According to the change, the relevant databases are updated as detailed in the "Decision Tree" and the "Cleaning Validation—Change Control Form". If the equipment and/or product grouping are changed due to the change, a new grouping should be performed by the validation unit. If the parameters in the contamination limit calculation equation are changed, the MC values should be recalculated.

The validation unit has to evaluate, according to the "Decision Tree," the effect of the change on the status of the equipment cleaning validation and justify the activities that should be taken, such as revalidation. This decision has to be reported on the "Cleaning Validation—Change Control Form". The form has to be signed and dated and sent to Quality Assurance.

Quality Assurance should evaluate the decision which was suggested and sign and date the form. Only then the activities which have to be done can be initiated.

Documentation

"Cleaning Validation—Change Control Form". Attachment II.

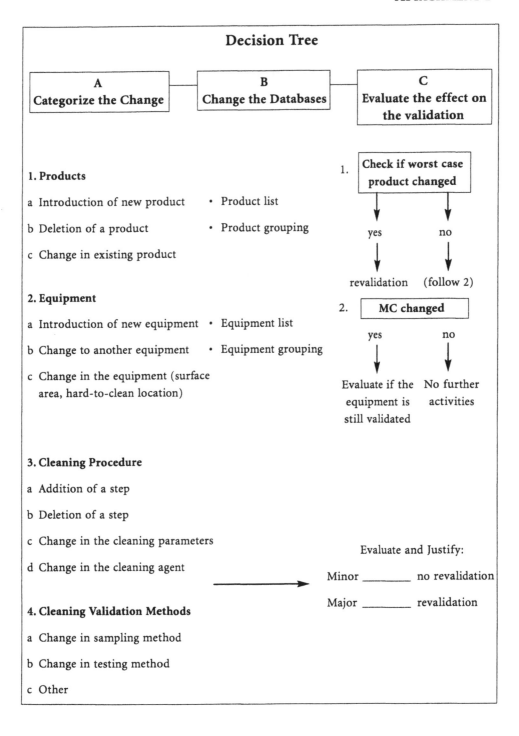

Change Control No.: _____

Cleaning Validation–Change Control Form

A. Report Change and Categorize

Change

❐ 1. Product Name: _____ Catalog No.: _____

❐ New   ❐ Deletion   ❐ Change in existing product   ❐ Other

❐ 2. Equipment Name: _____ Inventory No.: _____

❐ New   ❐ Deletion   ❐ Change in existing equipment   ❐ Other

❐ 3. Cleaning Procedure: _____ No.: _____

❐ New step   ❐ Deletion of step   ❐ Parameter: _____   ❐ Cleaning agent   ❐ Other

❐ 4. Cleaning Validation Method

❐ Sampling   ❐ Testing   ❐ Other

Detail the changes: _____
_____
_____

_____   _____
Signature   Date

B. Change the Databases

❐ 1. The following database is changed:

❐ Product list   ❐ Active ingredient list   ❐ Cleaning agent

❐ Equipment list   ❐ Equipment surface area   ❐ Cleaning procedure

Continued on next page

Continued from previous page

☐ 2. The following grouping is changed:

☐   ☐

Product grouping   Equipment grouping

☐ 3. The new MC values: ☐ active   _____

                                 ☐ cleaning agent _____

C. Evaluation of Change and Decision

_____

_____

_____

_____   _____

Validation unit   Quality Assurance

Signature/date   Signature/date

Work Instructions—Swab

SAMPLING FOR CLEANING VALIDATION

The Rx Pharmaceutical Company	WORK INSTRUCTIONS	
Edition No.:		Validation Unit:
Supersedes:	**Swab Sampling for Cleaning Validation**	Production:
Effective Date:		Quality Assurance:

page 1 of 2

Safety

Wear powder-free gloves and goggles.

Equipment and Materials

- Swab: Use approved swabs, such as Texwipe or Whatman paper.

- Solvent: For the active ingredient, use the solvent prescribed in the specific cleaning validation protocol. For the cleaning agent, use purified water USP/EP.

- Sampling frame: For flat surfaces, use a 5 cm × 5 cm stainless steel frame with a handle.

- Pincers

- Glass Erlenmeyer flasks

Procedure

Note: Extreme care should be taken to prevent swab contamination during handling.

Active ingredients:

1. Using clean pincers, soak the swab in the solvent and hold it until dripping stops.

2. Rub the saturated swab manually over the sample surface area designated in the equipment diagram appearing in the specific cleaning validation protocol. For flat surfaces, use the stainless steel frame. For smaller surfaces and hard-to-reach locations, rub the whole surface.

    Swabbing is performed in a controlled fashion by striking the sample surface area as indicated in the Figure 12.1.

The Rx Pharmaceutical Company		WORK INSTRUCTIONS
Edition No.:		Validation Unit:
Supersedes:	**Swab Sampling for Cleaning Validation**	Production:
Effective Date:		Quality Assurance:

3. Transfer the swab into a labeled glass Erlenmeyer flask and seal the flask.

4. Repeat the procedure for each location indicated on the equipment diagram and record the sampling on the sampling form of the cleaning validation protocol.

5. Submit samples for analysis.

6. Cleaning agents: Proceed as indicated for the active ingredient using purified water USP/EP.

Acceptance Criteria

Not applicable.

Documentation

Cleaning validation protocol sampling form.

Figure 12.1. Swab Sampling

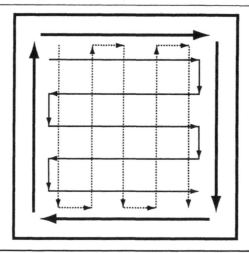

Work Instructions—Swab

CHARACTERIZATION FOR CLEANING VALIDATION

The Rx Pharmaceutical Company	WORK INSTRUCTIONS	
Edition No.:	**Swab Characterization for Cleaning Validation**	Validation Unit:
Supersedes:		Production:
Effective Date:		Quality Assurance:

page 1 of 2

Safety

- Wear powder-free gloves and goggles.

- Use a fume cupboard while working with organic solvents.

Equipment and Materials

- Swabs to be tested.

- Analytical balance equipped with a printing device.

- Analytical instrumentation.

- Mechanical shaker.

- Pincers.

- Glass Erlenmeyer flasks.

- Purified water USP/EP.

- Solvent to be used to saturate swab.

- Solvent to be used to extract the swab according to the specific cleaning validation protocol.

Procedure

Swab absorption capacity

1. Individually weigh 10 unused dry swabs.

2. Soak each swab in the solvent.

3. Using pincers, remove the swab from the solvent, hold it until dripping stops, and place it immediately on the balance pan.

4. Individually weigh the 10 saturated swabs and calculate the swab solvent absorption capacity.

5. Repeat the same procedure using purified water instead of the solvent and calculate the swab water absorption capacity.

The Rx Pharmaceutical Company		WORK INSTRUCTIONS
Edition No.:	**Swab Characterization for Cleaning Validation**	Validation Unit:
Supersedes:		Production:
Effective Date:		Quality Assurance:

Swab interference

1. Place one unused dry swab in Erlenmeyer flasks, each containing one of the following solvents:

   • Purified water USP/EP

   • Swab saturating solvent

   • Swab extracting solvent to be used in the analytical test method

2. Shake for 10 minutes at 150 strokes/minute.

3. Repeat the procedure with 4 additional swabs for each solvent.

4. Test the extraction solvent by the analytical methods to be used to determine the active ingredient and the cleaning agent, respectively, and calculate the swab interference response, to be taken into account in the analytical determination.

Acceptance criteria

The swab suitability is based on the following considerations:

   • No change in swab appearance and color.

   • Swab durability without signs of disintegration, fibers, or sediments.

   • Swab solvents and water absorption capacity is not less than the volume of solvent or water needed to dissolve the calculated amount of residual active ingredient and cleaning agent, respectively.

Documentation

Swab Characterization Report, to be attached to the cleaning validation protocol.

Work Instructions—Sampling

RECOVERY TEST

The Rx Pharmaceutical Company	WORK INSTRUCTIONS	
Edition No.:		Validation Unit:
Supersedes:	**Sampling Recovery Test**	Production:
Effective Date:		Quality Assurance:

Safety

- Wear powder-free gloves and goggles.

- Use a fume cupboard while working with organic solvents.

Equipment and Materials

Equipment

- Stainless steel surface 10 cm × 10 cm or coupon made of the same material as the equipment construction material which should be sampled

- Swab used as the sampled location in the equipment: Name _____

- Swab used for cleaning agents sampling: Name _____

- Suitable glass Erlenmeyer flasks.

- Powder-free gloves

- Clean pincers

Materials

- Extracting solvent for active material sampling:
  Choose the appropriate solvent taking into consideration the solubility of the contaminant active ingredient(s) in it. The toxicity of the solvent should also be considered. The most useful solvent is ethanol:

- Purified water USP/EP for cleaning agents sampling.

- Active ingredient(s) of the product to be validated for cleaning.

- Cleaning agents used for the relevant cleaning procedure.

The Rx Pharmaceutical Company	WORK INSTRUCTIONS	
Edition No.:		Validation Unit:
Supersedes:	**Sampling Recovery Test**	Production:
Effective Date:		Quality Assurance:

Procedure

The aim of this test is to determine the reliability of the sampling method, and to take into account loss due to sampling limitation.

This test is to be performed for the predetermined active ingredient(s) and cleaning agent(s) in accordance with the specific protocol for the equipment.

Method for Active Materials

1. Using the appropriate solvent, prepare a stock solution/dispersion calculated to yield the approximate amount of the active material as calculated for the maximal allowable contamination level (but not more than × 3 of the calculated value) per 0.5 mL.

2. Apply a measured amount of the stock dispersion/solution to the test surface (10 × 10 cm) and dry it in an oven (37°C).

3. Allow the sample to cool to room temperature.

4. Wear gloves and take at least 2 swab samples (using new gloves and a different swab wetted with the extracting solvent) from the test surface.

5. Transfer each swab into a suitable labeled glass Erlenmeyer flask and protect it from external contamination. The sampling must be completed as quickly as possible to prevent solvent evaporation and contaminant redeposition on the sampled surface.

6. Submit the labeled sealed glass Erlenmeyer flask to the analytical laboratory immediately along with the blank sample (one swab wetted with the extracting solvent) and the appropriate analytical request form.

7. The sampling recovery value is the ratio between the measured values and the expected values.

Method for cleaning agents

This test will be performed for the cleaning agents being used in the equipment cleaning procedures.

1. List all cleaning agents used to clean the equipment.

The Rx Pharmaceutical Company	WORK INSTRUCTIONS	
Edition No.:		Validation Unit:
Supersedes:	**Sampling Recovery Test**	Production:
Effective Date:		Quality Assurance:

2. Using purified water USP/EP, prepare a stock solution calculated to yield the approximate amount of cleaning agent as calculated for the maximal allowable contamination level (but not more than × 3 of the calculated value).

3. Apply a measured amount of the stock solution to the test surface (10 × 10 cm) and dry it in an oven (37°C).

4. Wear gloves and take at least two swab samples (using new gloves and different swab wetted with purified water USP/EP) from the test frame.

5. Apply each wipe into a suitable labeled glass Erlenmeyer flask.

6. Submit the labeled sealed glass Erlenmeyer flask to the analytical laboratory along with the blank sample (one swab wetted with purified water USP/EP) and the appropriate analytical request form.

7. The sampling recovery value will be the ratio between the measured values and the expected values.

Acceptance Criteria

- The overall sampling yield value of swabs should be not less than 70%.

- In cases where low results are obtained in a reproducible manner, the sample surface area may be sampled again using a second swab and the results obtained from both swabs summed up.

Documentation

- Request for Laboratory Control

- Sampling Recovery Study Report to be attached to the cleaning validation protocol.

Protocol and Report

CLEANING VALIDATION

The Rx Pharmaceutical Company		
Edition No.:	**Cleaning Validation Protocol & Report**	Validation Unit:
Supersedes:		Production:
Effective Date:		Quality Assurance:

page 1 of 13

Equipment Group No. Equipment	
Inventory No. Equipment Name	_____ _____

Contents:

• Approval Sheet

• Protocol/Report

• Attachments

Reference: Cleaning Validation Master Plan, Edition no. ____, dated _____

Date: _____

APPROVAL SHEET

The Rx Pharmaceutical Company		
Edition No.:	Cleaning Validation Protocol & Report	Validation Unit:
Supersedes:		Production:
Effective Date:		Quality Assurance:

page 2 of 13

This study has been performed according the cleaning validation protocol by:

Name _____   Signature _____   Date _____

and has been reviewed and approved by:

Name _____   Signature _____   Date _____
        Validation unit

Name _____   Signature _____   Date _____
        Production

Name _____   Signature _____   Date _____
        Quality assurance

PROTOCOL

The Rx Pharmaceutical Company		
Edition No.:		Validation Unit:
Supersedes:	**Cleaning Validation Protocol & Report**	Production:
Effective Date:		Quality Assurance:

page 3 of 13

Purpose

The purpose of this cleaning validation protocol is to demonstrate that the cleaning procedure used to clean Equipment no. X, selected as the worst-case equipment of Group no. Y, can effectively remove residues of the products manufactured in Equipment X and of the cleaning agent K used to clean Equipment X, to a predetermined level that does not raise patient safety concerns.

Responsibility

Validation unit:

- Prepare the cleaning validation protocol
- Conduct the validation study, including sampling
- Prepare and approve the cleaning validation report

Production:

- Approve the cleaning validation protocol
- Perform cleaning
- Approve the cleaning validation report

Quality assurance:

- Approve the cleaning validation protocol
- Approve the cleaning validation report

	The Rx Pharmaceutical Company	
Edition No.:		Validation Unit:
Supersedes:	**Cleaning Validation Protocol & Report**	Production:
Effective Date:		Quality Assurance:

Frequency

The cleaning validation study is performed after three consecutive major cleanings of Equipment X after the manufacture of Product A which contains the most insoluble of the active ingredients of the group of products manufactured in Equipment X.

Procedure

Follow the instructions of SOP No. _____ Cleaning Validation. Prepare, review, and attach the following documents:

1. Equipment:

   • Equipment database ❐

   • Equipment drawings and sampling locations ❐

2. Product and cleaning agent databases ❐

3. Calculation of the contamination limits ❐

4. Visual inspection and sampling ❐

5. Analytical results ❐

6. Analytical studies ❐

7. Conclusion ❐

Acceptance Criteria ❐

Contamination limits

Documentation

The cleaning validation report consists of the filled-in cleaning validation protocol and attachments.

The Rx Pharmaceutical Company		
Edition No.:	**Cleaning Validation Protocol & Report**	Validation Unit:
Supersedes:		Production:
Effective Date:		Quality Assurance:

page 5 of 13

EQUIPMENT DATABASE

List of Equipment:

Group No.	Equipment No.	Equipment Name	Operating Principle	Product Contact Surface Area	Hard-to-Clean Locations
1					
2					
3					
4					
5					
6					
7					
8					

Selected Equipment Group and Worst-Case Equipment:

Cleaning Procedure:

No. _____ Edition no. _____ Verified by _____ Date _____

Signature _____   Date _____

	The Rx Pharmaceutical Company	
Edition No.:		Validation Unit:
Supersedes:	**Cleaning Validation Protocol & Report**	Production:
Effective Date:		Quality Assurance:

EQUIPMENT DRAWING AND SAMPLING LOCATIONS

Equipment Part Drawings

1.

2.

3.

	The Rx Pharmaceutical Company		
Edition No.:		Validation Unit:	
Supersedes:	**Cleaning Validation Protocol & Report**	Production:	
Effective Date:		Quality Assurance:	

EQUIPMENT PART DRAWING (CONTINUED)

4.

5.

Sampling Locations and Swab Sampling Areas

	Part Name	Part Surface Area (cm^2)	Active Ingredient		Cleaning Agent	
			Sampling Location	Swab Area Ss (cm^2)	Sampling Location	Swab Area Ss (cm^2)
1						
2						
3						
4						
5						

Signature_____   Date _____

The Rx Pharmaceutical Company		
Edition No.:	**Cleaning Validation Protocol & Report**	Validation Unit:
Supersedes:		Production:
Effective Date:		Quality Assurance:

PRODUCT AND CLEANING AGENT DATABASE

List of Products

Group No.	Equipment No.	Equipment Name	Product Name	Active Ingredient	Solubility In water (mg/ml)	LTD (mg)	Daily Dose D (units/day)	Unit Dose Weight Wt (g)	Batch Size Wb (g)
1									
2									
3									

List of Cleaning Agents

Group No.	Equipment No.	Equipment Name	Cleaning Agent	Active Ingredient	Solubility In water (mg/ml)	LTD (mg)	Virtual Product Daily Dose D (units/day)	Virtual Product Unit Weight Wt (g)	Virtual Product Batch Size Wb (g)
1									
2									
3									

Selected Worst Case Product:

_____

Selected Worst Case Cleaning Agent:

_____

Signature _____   Date _____

	The Rx Pharmaceutical Company	
Edition No.:		Validation Unit:
Supersedes:	**Cleaning Validation Protocol & Report**	Production:
Effective Date:		Quality Assurance:

CALCULATION OF THE CONTAMINATION LIMIT

Active Ingredient

$$MC \text{ (mg/swab)} = \frac{LTD/1000}{D} \times \frac{Wb}{Wt} \times \frac{Ss}{Se} \times R$$

$$MC \text{ (mg/swab)} = \underline{\hspace{3cm}}$$
$$\text{(active)}$$

Cleaning Agent

$$MC \text{ (mg/swab)} = \frac{LD50/1000}{D} \times \frac{Wb}{Wt} \times \frac{Ss}{Se} \times R$$

$$MC \text{ (mg/swab)} = \underline{\hspace{3cm}}$$
$$\text{(cleaning agent)}$$

Signature _____   Date _____

The Rx Pharmaceutical Company		
Edition No.:	**Cleaning Validation Protocol & Report**	Validation Unit:
Supersedes:		Production:
Effective Date:		Quality Assurance:

VISUAL INSPECTION AND SAMPLING

Product Name: _____

Batch No.: _____

Part Name	Visual Inspection		Sampled for	
	Clean	Unclean	Active Ingredient	Cleaning Agent
1.	☐	☐	☐	☐
2.	☐	☐	☐	☐
3.	☐	☐	☐	☐
4.	☐	☐	☐	☐
5.	☐	☐	☐	☐

Remarks: _____

Signature: _____ Date: _____

Product Name: _____

Batch No.: _____

Part Name	Visual Inspection		Sampled for	
	Clean	Unclean	Active Ingredient	Cleaning Agent
1.	☐	☐	☐	☐
2.	☐	☐	☐	☐
3.	☐	☐	☐	☐
4.	☐	☐	☐	☐
5.	☐	☐	☐	☐

Remarks:

Signature: _____ Date: _____

Continued on next page

Continued from previous page

Product Name: _____

Batch No.: _____

Part Name	Visual Inspection		Sampled for	
	Clean	Unclean	Active Ingredient	Cleaning Agent
1.	❏	❏	❏	❏
2.	❏	❏	❏	❏
3.	❏	❏	❏	❏
4.	❏	❏	❏	❏
5.	❏	❏	❏	❏

Remarks: _____

Signature: _____   Date: _____

	The Rx Pharmaceutical Company	
Edition No.:		Validation Unit:
Supersedes:	**Cleaning Validation Protocol & Report**	Production:
Effective Date:		Quality Assurance:

ANALYTICAL RESULTS

Product Name: **Batch No.**

	Equipment Part	Active Ingredient (mg/swab)	Pass	Fail	Cleaning Agent (mg/swab)	Pass	Fail
1			☐	☐		☐	☐
2			☐	☐		☐	☐
3			☐	☐		☐	☐
4			☐	☐		☐	☐
5			☐	☐		☐	☐

Product Name: **Batch No.**

	Equipment Part	Active Ingredient (mg/swab)	Pass	Fail	Cleaning Agent (mg/swab)	Pass	Fail
1			☐	☐		☐	☐
2			☐	☐		☐	☐
3			☐	☐		☐	☐
4			☐	☐		☐	☐
5			☐	☐		☐	☐

Continued on next page

Continued from previous page

Product Name: **Batch No.**

	Equipment Part	Active Ingredient (mg/swab)	Pass	Fail	Cleaning Agent (mg/swab)	Pass	Fail
1			☐	☐		☐	☐
2			☐	☐		☐	☐
3			☐	☐		☐	☐
4			☐	☐		☐	☐
5			☐	☐		☐	☐

Signature _____   Date _____

The Rx Pharmaceutical Company		
Edition No.:		Validation Unit:
Supersedes:	**Cleaning Validation Protocol & Report**	Production:
Effective Date:		Quality Assurance:

ANALYTICAL STUDIES

Attach the following documents:

Analytical Report No. _____   Date _____

Analytical Method

	Active ingredient	Cleaning agent
Analytical Method No.		
Edition No.		
Method Validation		
Edition No.		
Detection Limit (DL)		
Quantitation Limit (QL)		

Recovery study

	% Recovery Active Ingredient	% Recovery Cleaning Agent
Recovery Study No. _____ Date _____		

Swab Characterization Study

Study Report	No.	Date

Signature _____   Date _____

The Rx Pharmaceutical Company		
Edition No.:	**Cleaning Validation Protocol & Report**	Validation Unit:
Supersedes:		Production:
Effective Date:		Quality Assurance:

CONCLUSION

The results obtained do not exceed the calculated contamination limit. The cleaning procedure of all pieces of equipment in Equipment Group no. X effectively removes residues of all the products manufactured in this equipment and is therefore deemed validated.

References

Agalloco, J., "Points to Consider in the Validation of Equipment Cleaning Procedures", *PDA Journal of Parenteral Science & Technology* 46(5), (1992): 163–168.

Allan, W., and T. Deeks, "Philosophy and Validation Approach to Cleaning and Decontamination in Antibiotics Facilities", *European Journal of Parenteral Sciences* 1(4), (1996): 107–112.

Baffi, R., G. Dolch, R. Garnick, Y. F. Huang, B. Mar, D. Matsuhiro, B. Niepelt, C. Parra, and M. Stephan, "A Total Organic Carbon Analysis Method for Validation Cleaning Between Products in Biopharmaceutical Manufacturing", *PDA Journal of Parenteral Science & Technology* 45 (1), (1991): 13–19.

Baseman, H. J., "SIP/CIP Validation", *Pharmaceutical Engineering* 12(2), (1992): 37–46.

Conine, D. L., B. D. Naumann, and L. H. Hecker, "Setting Health-Based Residue Limits for Contaminants in Pharmaceuticals and Medical Devices", *Quality Assurance: Good Practice, Regulatory and Law* 1(3), (1992): 171–180.

Cooper, D. W., "Cleaning Validation and Monitoring Aseptic Fill Areas", *Pharmaceutical Technology Europe* (January 1998): 31–35.

Dictionnaire Vidal, France, Current Edition.

Dourson, M. L., and J. F. Stara, "Regulatory History and Experimental Support of Uncertainty (Safety) Factors", *Regulatory Toxicology and Pharmacology* 3 (1983): 224–238.

Drug GMP Report No. 59 (June 1997), Washington Business Information Inc., Arlington, VA.

Drug GMP Report No. 71 (June 1998), Washington Business Information Inc., Arlington, VA.

EC Guide to Good Manufacturing Practice for Medicinal Products, 1997.

Faubel, G., W. Schott, and C. van der Stouwe, "The Problem of Cleaning Pharmaceutical Packaging Machines for Solid Oral Formulations", *Drugs Made in Germany* 33(2), (1990): 52–96.

FDA, *FDA Guide to Inspection of Bulk Pharmaceutical Chemicals* (Washington, D.C.: FDA, revised September 1991).

FDA, FDA CGMP Regulations-Decision in *United States vs. Barr Laboratories, Inc.* by the U.S. District Court for the District of New Jersey, USA (February 1993).

FDA, *FDA Mid Atlantic Region Inspection Guide,* "Cleaning Validation" (Washington, D.C.: FDA, revised May 7, 1993).

FDA, *FDA Guide to Inspections of Validation of Cleaning Processes* (Washington D.C.: FDA, July 1993).

FDA, *FDA Guide to Inspections of Water Systems* (Washington D.C.: FDA, 1993)

FDA, *FDA Validation Documentation Inspection Guide* (Washington, D.C.: FDA, 1993).

FDA, *FDA Draft Guidance for Industry Manufacturing, Processing, or Holding Active Pharmaceutical Ingredients* (Washington, D.C.: FDA, March 1998).

FDA, *Code of Federal Regulations of the FDA* 21 CFR, Part 211. (Washington, D.C.: FDA, revised April 1999).

FDA, FDA Guidance for Industry, *SUPAC-IR/MR: Immediate Release and Modified Release Solid Oral Dosage Forms Manufacturing Equipment Addendum* (Washington D.C.: FDA, 1999).

Flickinger, B., "Charting a Course for Cleaning Validation", *Pharmaceutical & Cosmetic Quality* (January/February 1997): 18–23.

Forsyth, R. J., and D. V. Haynes, "Cleaning Validation in a Pharmaceutical Research Facility", *Pharmaceutical Technology Europe* 19–24 (January 1999).

Fourman, G. L., and M. V. Mullen, "Determining Cleaning Validation Acceptance Limits for Pharmaceutical Manufacturing Operations", *Pharmaceutical Technology* (April 1993): 54–60. *Pharmaceutical Technology International* (June 1993): 46–49.

Gavlick, W. K., L. A. Ohlemeier, and H. J. Kaiser; "Analytical Strategies for Cleaning Agent Residue Determination", *Pharmaceutical Technology* (March 1995): 136–144.

"The Gold Sheet" Vol. 27, No. 5 (May 1993) F-D-C Reports Inc., Chevy Chase, MD.

"The Gold Sheet" Vol. 32, No. 2 (February 1998) F-D-C Reports Inc., Chevy Chase, MD.

"GMP Trends", Issue No. 479 (January 1, 1997), GMP Trends Inc., Boulder, CO.

"GMP Trends", Issue No. 500 (November 15, 1997), GMP Trends Inc., Boulder, CO.

"GMP Trends", Issue No. 501 (December 1, 1997), GMP Trends Inc., Boulder, CO.

"GMP Trends", Issue No. XXVII (June 1998), GMP Trends Inc., Boulder, CO.

Hakimipour, F., "Penicillin Decontamination Procedures for a Pharmaceutical Manufacturing Facility", *Pharmaceutical Technology* (June 1984): 88–94.

Harder, S. W., "The Validation of Cleaning Procedures", *Pharmaceutical Technology* (May 1984): 29–34.

Hausmann, R., J. J. Kaufmann, and K. Richter, "Tabletting Technology Including a CIP System and Die-Filling by Centrifugal Force", *Pharmaceutical Technology Europe* (April 1998): 18–24.

Holmes, A. J., and A. J. Vanderwielen, "Total Organic Carbon Method for Aspirin Cleaning Validation", *PDA Journal of Pharmaceutical Science and Technology* 51(4), (1997): 149–152.

Human Drug cGMP Notes (June 1998). FDA, Office of Compliance, Center for Drug Evaluation and Research, Washington D.C.

Hwang, R. C., D. L. Kowalski, and J. E. Truelove, "Process Design and Data Analysis for Cleaning Validation", *Pharmaceutical Technology* (January 1997): 62–68. *Pharmaceutical Technology Europe* (February 1997): 21–25.

Inspection Monitor, Vol. III, No. 6 (June 1998), Washington Information Source Co., Rockville, MD, USA.

International Conference on Harmonization (ICH), Q2A "Validation of Analytical Procedures" (March 1995).

International Conference on Harmonization (ICH), Q2B "Validation of Analytical Procedures: Methodology" (May 1997).

Jenkins, K. M., and A. J. Vanderwielen, "Cleaning Validation: An Overall Perspective", *Pharmaceutical Technology* (April 1994): 60–73.

Jenkins, K. M., A. J. Vanderwielen, J. A. Armstrong, L. M. Leonard, G. P. Murphy, and N. A. Piros, "Application of Total Organic Carbon Analysis to Cleaning Validation", *PDA Journal of Pharmaceutical Science & Technology* 50(1), (1996): 6–15.

Kieffer, R. G., "Validation and the Human Element," *PDA Journal of Pharmaceutical Science & Technology* 52(2), (March/April 1998): 52–54.

Kirsch, R. B., "Validation of Analytical Methods Used in Pharmaceutical Cleaning Assessment and Validation" *Pharmaceutical Technology Analytical Validation* (1998): 40–46.

Laban, F., M. Cauwet, V. Champault, P. R. Dampfhoffer, E. Delestre, S. Detoc, F. Durand, M. J. Girault, L. Grillet, A. Loret, C. Martin-Delory, P. Michel, C. Nivet, A. Picaut, E. Prevost, M. Sarradin, R. de la Tour, P. Trotemann, and J. Willems, "Validation of Cleaning Procedures Report of an SFSTP Commission", S.T.P. *Pharma Pratiques* 7(2), (1997): 87–127.

Layton, D. W., B. J. Mallon, D. H. Rosenblatt, and M. J. Small; "Deriving Allowable Daily Intakes for Systemic Toxicants Lacking Chronic Toxicity Data", *Regulatory Toxicology and Pharmacology* 7 (1987): 96–112.

Lazar, M., "PhRMA Guidelines for the Validation of Cleaning Procedures for Bulk Pharmaceutical Chemicals", *Pharmaceutical Technology* (September 1997): 56–73.

LeBlanc, D. A., "Sampling, Analyzing, and Removing Surface Residues Found in Pharmaceutical Manufacturing Equipment" *Microcontamination* (May 1993): 37–40.

LeBlanc, D. A., "Rinse Sampling for Cleaning Validation Studies", *Pharmaceutical Technology* (May 1998): 67–74.

LeBlanc, D. A., "Establishing Scientifically Justified Acceptance Criteria for Cleaning Validation of Finished Drug Products", *Pharmaceutical Technology* (October 1998): 136–148.

LeBlanc, D. A., D. D. Danforth, and J. M. Smith, "Cleaning Technology for Pharmaceutical Manufacturing", *Pharmaceutical Technology* (July 1993): 84–92.

Lombardo, S., P. Inampudi, A. Scotton, G. Ruezinsky, R. Rupp, and S. Nigam, "Development of Surface Swabbing Procedures for a Cleaning Validation Program in a Biopharmaceutical Manufacturing Facility" *Biotechnology and Bioengineering* 48 (5), (1995): 513–519.

Martindale, *The Extra Pharmacopoeia*, Royal Pharmaceutical Society, London U.K. Current Edition.

McArthur, P. R., and M. Vasilevsky, "Cleaning Validation", *Proceedings—Institute of Environmental Sciences* 41 (1995): 302–309.

McArthur, P. R., and M. Vasilevsky, "Cleaning Validation for Biological Products: A Case Study", *Pharmaceutical Engineering* (November/December 1995): 24–31.

McCormick, P. Y., and L. F. Cullen, "Cleaning Validation", in *Pharmaceutical Process Validation* 2nd ed. New York: Marcel Dekker, 1993.

Mendenhall, D. W., "Cleaning Validation." *Drug Development and Industrial Pharmacy* 15 (13), (1989): 2105–2114.

The Merck Index, Merck and Co. Inc., Rahway, NJ, Current Edition.

Mirza, T., R. C. George, J. R. Bodenmiller, and S. A. Belanich, "Capillary Gas Chromatographic Assay of Residual Methanamine Hippurate in Equipment Cleaning Validation Swabs", *Journal of Pharmaceutical and Biomedical Analysis* 16 (1998): 939–950.

Nilsen, C. L., "A Sensible Plan for the Design and Execution of Cleaning Validations" in *The QC Laboratory Chemist: Plain and Simple*, Buffalo Grove, IL: Interpharm Press, Inc., 1998.

Parikh, D. M., "Current Issues and Troubleshooting Fluid Bed Granulation", *Pharmaceutical Technology Europe* (May 1998): 42–48.

PDA Technical Report No. 29, Points to Consider for Cleaning Validation (November/December 1998, Supplement), *PDA Journal of Pharmaceutical Science and Technology* vol. 52, No. 6.

Physicians Desk Reference, Current Edition. Medical Economics Company, Inc., Montvale, NJ.

PIC-PIC/S, Draft "Internationally Harmonized Guide for Active Pharmaceutical Ingredients, Good Manufacturing Practice", Convention for the Mutual Recognition of Inspections in Respect of the Manufacture of Pharmaceutical Products (September 1997).

"Principles of Qualification and Validation in Pharmaceutical Manufacture", Document PH Convention for the Mutual Recognition of Inspections in Respect of the Manufacture of Pharmaceutical Products (1999).

Romanach, R. J., S. F. Garcia, O. Villanueva, and F. Perez, "Combining Efforts to Clean Equipment in Active Pharmaceutical Ingredient Facilities", *Pharmaceutical Technology* (January 1999): 46–58.

Rote Liste, Germany, Current Edition.

Shea, J. A., W. F. Shamrock, C. A. Abboud, R. W. Woodeshick, L. Q. S. Nguyen, J. T. Rubino, and J. Segretario "Validation of Cleaning Procedures for Highly Potent Drugs. I. Losoxantrone", *Pharmaceutical Development and Technology* 1(1), (1996): 69–75.

Smith, J. M., "A Modified Swabbing Technique for Validation of Detergent Residues in Clean-in-Place Systems", *Pharmaceutical Technology* (January 1992): 60–66.

Smith, J. M., "Selecting Analytical Methods to Detect Residue from Cleaning Compounds in Validated Process Systems", *Pharmaceutical Technology* (June 1993): 88–98.

Sofer, G. "Validation of Biotechnology Products and Processes", *Current Opinion in Biotechnology* 6 (1995): 230–234.

U.S. Pharmacopeia 23/National Formulatory 18, 1995 United States Pharmacopeial Convention, Inc. Rockville, MD, USA.

Washington Drug Letter 27 No. 9, Washington Business Information Inc., Arlington, VA. (February 27, 1995).

Washington Drug Letter 27 No. 12, Washington Business Information Inc., Arlington, VA. (March 27, 1995).

Washington Drug Letter 28 No. 5, Washington Business Information Inc., Arlington, VA. (January 29, 1996).

Washington Drug Letter 30 No. 24 Washington Business Information Inc., Arlington, VA. (June 15, 1998).

WHO Good Manufacturing Practices for Pharmaceutical Products, 1992.

Zeller, A. O., "Cleaning Validation and Residue Limits: A Contribution to Current Discussions", *Pharmaceutical Technology* (October 1993): 70–80.

Index